THE ROHONC CODE

THE ROHONC CODE

Tracing a Historical Riddle

Benedek Láng

Translated by Benedek Láng,
Teodóra Király, and Nick Palmer

THE PENNSYLVANIA STATE UNIVERSITY PRESS

UNIVERSITY PARK, PENNSYLVANIA

Library of Congress Cataloging-in-Publication Data

Names: Láng, Benedek, 1974– author, translator. | Király,
    Teodóra, translator. | Palmer, Nick (Translator), translator.
Title: The Rohonc code : tracing a historical riddle / Benedek
    Láng ; translated by Benedek Láng, Teodóra Király, and
    Nick Palmer.
Description: University Park, Pennsylvania : The Pennsylvania
    State University Press, [2021] | Includes bibliographical
    references and index.
Summary: "An investigation into the history of ciphers,
    codebreaking, and artificial languages that focuses on
    the Rohonc Codex, a four-hundred-year-old manuscript
    written entirely in cipher"—Provided by publisher.
Identifiers: LCCN 2021001997 | ISBN 9780271090207
    (paperback)
Subjects: LCSH: Rohonci kódex. | Cryptography—Hungary.
    | Ciphers.
Classification: LCC Z103.4.H86 L3613 2021 | DDC
    652/.809439—dc23
LC record available at https://lccn.loc.gov/2021001997
Parts of this book have appeared in print in the following
publications: "Why Don't We Decipher an Outdated Cipher
System? The Codex of Rohonc," *Cryptologia* 34 (2010): 115–44;
"People's Secrets: Towards a Social History of Early Modern
Cryptography," *Sixteenth Century Journal* 45, no. 2 (2014):
291–308; "Invented Middle Ages in Nineteenth-Century
Hungary: The Forgeries of Sámuel Literáti Nemes," in
*Manufacturing a Past for the Present: Forgery and Authenticity
in Medievalist Texts and Objects in Nineteenth-Century
Europe*, ed. Patrick Geary and Klaniczay Gábor (Leiden: Brill,
2015), 129–43; and *Real Life Cryptology: Ciphers and Secrets in
Early Modern Hungary* (Amsterdam: Amsterdam University
Press, 2018), 31–49.

CONTENTS

List of Illustrations | vi
Acknowledgments | vii

Introduction | 1

1. The Treasure-Hunting Bookdealer | 8

2. Codebreaking Attempts | 23

3. The Quest Begins | 35

4. The Jesus Code | 42

5. Writing in Ciphers and Codes | 62

6. Decryption | 80

7. Shorthand Systems | 97

8. Artificial Languages and Codes | 102

9. Methods of Codebreaking | 119

Coda | 132

Appendix | 137
Notes | 145
Bibliography | 153
Index | 163

ILLUSTRATIONS

FIGURES

1. Two pages of the Rohonc Codex | 5
2. The table of characters of the Gelle prayer book | 20
3. The first page of the Gelle prayer book | 21
4. Unidentified scene, probably Moses with the Tablets of Law on Mount Sinai | 44
5. Christ entering Jerusalem and the cleansing of the Temple | 44
6. The Annunciation; Joseph and the angel; a sacrifice in the Temple | 45
7. The Adoration of the Magi | 45
8. The Transfiguration of Jesus or a baptism scene | 46
9. The Crucifixion | 46
10. Christ standing before Pilate | 47
11. Christ before Pilate wearing the crown of thorns | 47
12. Two altars, each with a priest, at the foot of the cross | 48
13. The events following the death of Christ: the *Noli me tangere* scene; the empty tomb with the angel; and John at the tomb | 48
14. The Resurrection of Christ | 48
15. Apocalyptic scene; the Tree of Knowledge, the four rivers, and Christ standing between two people | 49
16. The structure of the world with the spheres and the heavenly bodies | 50
17. A crowned person praying to the sun | 50
18. The monoalphabetic cipher of Ludovicus de Berlaymont, archbishop of Cambrai | 63
19. Homophonic cipher from Milan, 1448 | 71
20. The Lord's Prayer in the language devised by John Wilkins | 107
21. The character language of György Kalmár | 110
22. The pictographic design of István Kovácsházi, illustrated with the Lord's Prayer | 113
23. Levente Zoltán Király's comparative study of the most frequent repetitions in the Rohonc Codex | 126

TABLE

1. A modern syllable-coding cipher | 95

ACKNOWLEDGMENTS

I would like to thank the following scholars, friends, and scholar-friends for their help: Balázs Ablonczy, Gábor Almási, Péter Banyó, Craig Bauer, Dóra Bobory, Gergely Buzás, Elonka Dunin, Claire Fanger, Eduard Frunzeanu, Ottó Gecser, Levente Zoltán Király, Gábor Klaniczay, József Laszlovszky, Beáta Megyesi, István Monok, János Németh, Joe Nickell, Veronika Novák, Nick Pelling, Jolanta Rzegocka, Dóra Sallay, Klaus Schmeh, Géza Szentmártoni Szabó, Gábor Tokai, László Ulicska; members of the Department of Philosophy and History of Science at the Budapest University of Technology and Economics Márta Fehér, Tihamér Margitay, Viktor Binzberger, János Tanács, and Gábor Zemplén; my parents, Zsuzsanna Láng and Csaba Láng; and finally my wife, Márta Tarnai, who learned much more about the Rohonc Codex than she originally wished to.

I am furthermore grateful to the Maison des Sciences de l'Homme in Paris, where I was able to concentrate exclusively on the codex thanks to a Mellon Fellowship; to the Hungarian National grant OTKA K 101544; and to the Swedish Research Council, grant 2018–06074, DECRYPT, Decryption of Historical Manuscripts, which supported the final stages of this work. Finally, I thank Richárd Rados, director of the Jaffa Publishing House, which published an earlier, Hungarian version of this book.

INTRODUCTION

NEVER-READ BOOKS

The success of an investigation depends mostly on how skillfully you can make objects talk. Things are born silent and incapable of talking to their inspector, though it would be most desirable if a weapon, for example, could testify against the criminal, the drug could point out its dealer, or money could tell whether its source is corrupt. How convenient that would be!

Books are also born silent. This poses no problems as long as we know their author, who can tell us about the origins, background, content, and purpose of the book. Problems arise when the author is not known. The first decades of the seventeenth century saw brief manifestos published in German spreading information about the mysterious Rosicrucian Brotherhood. Their authors remained anonymous, and nobody could decide whether the powerful Rosicrucian group, which had allegedly been formed to benefit mankind and to reform science, even existed. Yet at least the content of these documents could be identified, even if the authors themselves and their purposes remained secret.

But what if the content is also inaccessible? Think of the ancient writings whose characters, once familiar to highly developed cultures, can no longer be read by anyone today. What about more recent ciphers that were intentionally encrypted into a string of weird characters and numbers? Or consider the messages coming from space: could they be from intelligent creatures trying to communicate with us, or are they mere background noise of the universe?

Making sense of these texts requires the work of specialists, inspectors equipped with the tools of history, philology, mathematics, and information technology. Historians, linguists, mathematicians, and other experts, spread out in different parts of the world, work on such cases simultaneously but in isolation from one another, only connecting from time to time at conferences to share the results of their research.

Occasionally, it takes generations of scholars and decades of work to achieve a breakthrough, as with the cuneiform writing of Mesopotamia, the Egyptian hieroglyphs, the ancient Greek writing known as Linear B, and the Mayan hieroglyphs. When that breakthrough occurs, however, and the scrawl that survived on a few pieces of papyrus, stone, or clay can finally be read, it opens up whole new worlds to us, telling of extinct cultures, histories, and religions of peoples long forgotten.

Not that persevering historical investigations always yield results. Neither the Linear A writing of Knossos, nor the Phaistos Disk, nor the Rongorongo system of glyphs (written with shark teeth) has yet been decoded. More recent codes, like the enigmatic Voynich manuscript, also remain a mystery. Failure to solve such puzzles generates frustration, which is exacerbated by the nagging feeling that the undeciphered writing may in fact be totally meaningless. The specialist is haunted by the fear that she may be the victim of a hoax, a vicious bluff, and by the possibility that what she considers ancient writing may simply have served a decorative purpose. Understanding the intentions of the author is crucial to identifying the content: did he encode a meaningful message, or was he only trying to make his text look mysterious, and thus more valuable, in the eyes of contemporaries or generations to come? Trying to determine whether a string of characters is random or whether it shows a pattern, the historian reaches for the special methods of mathematics and statistics. But the undertaking is fraught with uncertainty. It would be much easier if the given stone, papyrus, or clay fragment could simply tell us who wrote on it, what it is, and when it was inscribed.

Still, even the most maddening investigations yield some results, however humble. In the early stages of decoding the Egyptian hieroglyphs or

Linear B, important conclusions were reached that served as stepping stones for the people who finally broke the code and gained renown. Although the Voynich manuscript, the Phaistos Disk, and the Beale ciphers remain unbroken (the last supposedly revealing the whereabouts of great treasure), a number of attempted solutions have already failed, which is an important milestone in itself. Most scholars working on an unsolved case gain a degree of satisfaction even when they are unable to find a complete solution. They feel that they have had a great time playing the inspector, and they appreciate what they have learned despite their failure to decipher the unknown text.

This book is about an investigation of a similar kind. It tells the story of a quest, staged in libraries across the world, to uncover the truth about another book and its content. It contains a surprisingly large number of symbols (or are they letters?) and amazingly few combinations (words?). It is richly decorated with images depicting what appears to be Christ in a Gospel-like story.

This investigation has been sometimes invigorating, sometimes frustrating. It has been costly. While pursuing it, I unintentionally became a specialist in a number of issues concerning the history of science, mainly the early modern history of ciphers and artificial languages. In the unlikely case that, tomorrow or next week, someone proves that this book is a hoax, I will not regret the time I have invested in this historical hunt. I have enjoyed virtually every minute of it, and I have learned a lot.

A NEGLECTED MANUSCRIPT

The Rohonc Codex is in a paradoxical situation: while privileged to be in the elegant company of the most famous hitherto unsolved writing systems (together with the Phaistos Disk and Linear A),[1] it does stand out from the crowd of other undeciphered writings in that very few professional code-breakers have studied it since it was first discovered in 1838.

The situation becomes particularly striking if we compare the codex with a similar enigma, the (in)famous Voynich manuscript. This book has kept armies of amateur codebreakers, information technologists, and historians on their toes ever since the Polish antiquarian and bookdealer Wilfrid Michael Voynich (1865–1930) bought it in the 1910s from the Jesuits at Villa

Mondragone outside Rome. The book, with its odd drawings of biology, astronomy, and bathing scenes and its incomprehensible structure of writing, has created a whole subculture of codebreakers, who regularly gather at conferences and communicate through special blogs and email lists.

The Rohonc Codex resembles the Voynich manuscript in several respects. Both were written in a mysterious code; both are fairly lengthy (and thus appear to offer themselves as easy prey to codebreakers); and both are equally likely to be ciphers, codes, artificial languages, or forged documents. If each should prove to be a forged document, we will still have to decide in both cases whether the text was forged in the sixteenth century, the nineteenth century, or somewhere in between. And once we have answered these questions, we still need to solve the puzzle of why someone made the effort to forge them in the first place.

The main thing that sets the two texts apart is the level of interest and enthusiasm surrounding them. The Voynich manuscript captured the attention of the best World War II codebreakers, including that of William Friedman (1891–1969), who broke dozens of Japanese military codes.[2] It has been studied by a number of expert historians and philologists. Half a dozen respectable monographs discuss the problems and history of the deciphering process, and numerous web pages and e-mail lists fire fans' enthusiasm.[3] The Rohonc Codex, by contrast, received only a few abortive codebreaking attempts before the year 2000.

What can be known about the codex with any degree of certainty? It consists of nearly 450 pages (see fig. 1). The first and last few dozen leaves were detached from the book, thus rendering the original order of these pages a challenge in itself. There is no title page. The pages are made of paper. The small pages (3.9 × 4.7 in., or roughly 10 × 12 cm) on average contain nine to fourteen lines of characters, and there are ninety-some images altogether. The fairly destroyed leather binding was attached to the pages in the nineteenth century and thus tells us nothing about the early history of the book.

What we know of its history dates back only to 1838, when it was incorporated into the collection of the Hungarian Academy of Sciences as part of the thirty-thousand-book library of the late Hungarian nobleman Gusztáv Batthyány.[4] Since we have no other information about its origin, the codex is named after the town of Rohonc (then a place in western Hungary, today called Rechnitz, in Austria), the Batthyány family seat. However, since the

FIG. 1 | Two pages of the Rohonc Codex. Library of the Hungarian Academy of Sciences, MS K 114, fols. 91r–90v.

Batthyánys amassed their book collection from a great array of sources over several centuries, there is no proof that the codex is either Hungarian or central European in origin.

Soon after it was discovered, the mysterious manuscript garnered considerable academic attention, but only as long as it could be considered a potentially valuable piece of old Hungarian writing. The initial enthusiasm soon died out, giving way to disappointment, skepticism, and suspicion. By the end of the nineteenth century, the academic world had decided to regard it as a forgery, and virtually no serious study was published on it until the turn of the twenty-first century.

THE AUTHOR MEETS THE MYSTERY

I stumbled upon the Rohonc Codex through a series of chance events. In 2006 I attended the annual International Congress on Medieval Studies in Kalamazoo, a city in the American Midwest. Every year, two to three thousand lectures are presented during the four or five days of this event and are attended by some four thousand medievalists altogether.

I attended a sizable book fair at this conference and noticed a book on the Voynich manuscript that seemed fairly reader-friendly. I didn't know much about this enigmatic manuscript, but a Canadian friend who did talked me into buying the book. I found it so intriguing that I couldn't put it down for the entire length of my nine-hour flight back to Frankfurt.[5]

A couple of weeks later, while on a family vacation in a small Hungarian village, I was again reading the book. A historian friend of mine happened to visit us there, and seeing me with the book on the Voynich manuscript, she casually mentioned that the library of the Batthyány family, her own particular research field, contains a similar enciphered manuscript that nobody knows anything about. She must have been surprised to hear, three months later, that I had spent virtually all of the time since her visit studying that manuscript, the Rohonc Codex.

The months that followed were a period of intense highs and lows: I was either totally devoted to breaking the code or simply reading up on ciphers in general. My previous research topic, medieval manuscripts on magic, had lost its appeal entirely. (To be honest, it only sounds exciting to laypeople. Scholars of magic know that their work is just as tedious as the job of studying, say, the "traditions of inheritance of the noblemen of a French county in the fifteenth century, especially the female heirs.") I got a taste of what it was like not only to be interested in a research topic but to be deeply committed to solving a puzzle. I was enthralled. My friends began to worry about me. I visited museums across western Europe, studying biblical scenes engraved in ivory or depicted in breviaries. I spent three months in Paris at the Sainte-Geneviève Library and the Bibliothèque nationale, looking for stenographies (i.e., shorthand systems), code symbols, and artificial-language schemes. In my studio apartment a few streets from Notre Dame, I spent many nights compiling statistical analyses and running vowel tests on the script, until the characters of the codex would begin to dance before my eyes. I became disillusioned at times, to the relief of my worried friends. I despaired of getting to the bottom of the codex, of determining once and for all whether it was a hoax. When my chances of success looked dim, I promised myself that I would at least write an article on my research, and I did just that. Then I wrote another article. Then a paper in an international journal on cryptology.[6] Finally, I wrote this book, describing the mystery surrounding the Rohonc Codex as well as the story of my investigation.

This book does not follow a linear chronology, because I worked in parallel on the different stages of the quest in libraries in Budapest, Chicago, Paris, and the Vatican. Instead, it presents my findings in a logical order. My first task was to rule out the possibility that the codex is a mere hoax, because this would render pointless any attempt to solve it. Second, I had to study earlier attempts to break the code in order to avoid the common mistake of self-appointed codebreakers who fail to consult the work of their predecessors and end up reinventing the wheel. One primary feature of the codex is that it is heavily illustrated, and its images might provide clues that could lead to a breakthrough; thus the third step was to analyze these illustrations. Historians and philologists are always looking for historical contexts and analogies for the sources they examine, and that is why the fourth task of this study is to look at how the Rohonc code fits into the matrix of ciphers, stenographies, and artificial languages of the past—these parallels might help identify a successful codebreaking method. Finally, I show how such mysterious strings of signs and symbols might be decoded according to the deciphering conventions of other ciphers and codes, and—in conclusion—I put forward a solution to the Rohonc Codex that appeared just as I was finishing my own research.

This book attempts to achieve a number of goals. First of all, I would like to share all there is to know about the Rohonc Codex at present. Second, I would like to illustrate the work of historians—their tools, methods, and research scenarios. Their lives are far less boring and dusty than the public imagines them to be, though not quite as thrilling as Dan Brown makes them out to be, either. Third, I hope to entertain the reader, which for me is a joyous task, for this is a gripping story of a puzzling codex and my attempt to solve it once and for all.

CHAPTER 1

THE TREASURE-HUNTING BOOKDEALER

AUTHENTIC OR FORGED?

When nineteenth-century Hungarian scholars realized that the Rohonc Codex was not an old Hungarian script and convinced themselves that it could thus be nothing but a forgery, they started looking for the forger. A man named Sámuel Literáti Nemes (1794–1842) seemed to be the optimal candidate for that role. A renowned book and antique dealer, Literáti was an unusual character with an almost dual personality, who, paradoxically, could also be described as a typical figure of his age. An active collector of books, Literáti was employed by a number of noble families, including the Batthyánys. Literáti saved many an old Hungarian script or literary relic from destruction during the turmoil of the Revolution of 1848. Alongside authentic antiques, however, he also sold a number of forged documents, perplexing contemporary scholars to no little extent.[1] This mixture of legitimate activity in the book trade and double-dealing chicanery can be seen in the two chapters dedicated to Literáti in a handbook on great Hungarian book collectors: one focuses on his life as an antique dealer, while the other depicts him as a swindler who created forgeries.[2] By way of contrast, other famous collectors and library owners (Johannes Vitéz, Janus Pannonius,

King Matthias Corvinus, Johannes Sambucus, and Ferenc Széchényi) are featured in only one chapter each in that book.

Literáti's dual nature was in line with the general trend of this period in Hungarian history. Catering to a growing sense of national pride, important literary relics were being discovered by busy antique dealers like Literáti. Meanwhile, pseudohistorical sources, antiques of dubious origin, made-up myths, and nonexistent writing systems fired initial enthusiasm, followed by fury and disappointment among specialists who were becoming more and more sensitive to correctness in philology and the history of language.

Interestingly, this age produced figures even more dubious and equivocal than Literáti. Kálmán Thaly (1839–1909), for example, was a leading historian of early modern history and poetry, and he was also the founding editor of the prestigious scholarly journal *Centuries (Századok)*. He became a member of the Hungarian Academy of Sciences fairly early in his long career. Thaly earned his reputation for being the pioneer researcher of Rákóczi's War of Independence—an eight-year, though finally aborted, uprising led by the Hungarians against the Habsburg dynasty in the first years of the eighteenth century. He spent much time on horseback and on wagons, traveling between the castles of Hungarian nobles, hunting down manuscripts in their private libraries. Curiously enough, he divided his time between studying these historical sources carefully and using his knowledge to *write* "historic" Hungarian poetry. He even had the audacity to forge a number of historic letters. The poems, called "Kuruc songs," were revealed as counterfeit only after the death of the celebrated historian-poet, but the authenticity of some of the other documents he left behind is still debated.[3]

The practice of rewriting old texts and re-creating the past in a way that goes far beyond just adding an archaic touch is not exclusive to Hungarian historians. One example is the famous Ossian poems, purported to be by the son of Fingal, king of Ireland, which were allegedly collected in the Scottish Highlands and then translated from Gaelic. These brought no small amount of money and fame to James Macpherson (1736–1796), the person who had "discovered" them. They were shown to be fake in the end, but only following the death of their author, the revered scholar.[4]

It was in this era that Sámuel Literáti Nemes began his career in Târgu Mureş (Marosvásárhely, a Hungarian town in Transylvania at the time, today in Romania). As a merchant, he traveled through the small villages of Transylvania and then as far as Graz, Zara, and Venice in his hunt for antiques.

After 1828, families of the nobility were increasingly keen on employing his services. Becoming a fashionable bookdealer, in 1835 he hung his shop sign, featuring a mammoth skeleton, over his antique shop in Buda, where he offered to sell, buy, or exchange "all kinds of antiques: money, weapons, ancient works, books on parchment, certificates, manuscripts, paintings and such."[5] Most of the books he collected were ultimately donated to the main Hungarian library, the National Széchényi Library in Budapest, so some regarded him as one of the founders of the collection (joining Ferenc Széchényi, of course, after whom the library is named.)

The forged documents created or sold by him, however, cast a shadow over the book collector's glory. He sold his first counterfeit, a sheet of "Chinese characters," in 1830, while the last forged documents were sold by his heirs after his death. Some of these were soon exposed by his customers, but a few of them enjoyed a longer life: first, Literáti's customer the linguist János Jerney appreciated the newly discovered sources in an enthusiastic article, which was followed by two decades of suspicion and silence, which in turn was ended when the philologist Károly Szabó debunked them in a long article in 1866.[6] Szabó relied on both codicology and historical linguistics: he called attention to the strange material in some of the manuscripts and pointed out that some texts contained word forms that had no equivalents in the time of their alleged origin; he was also suspicious about the ink and the form of some written characters, which reminded him more of the eighteenth century than of the fourteenth or fifteenth.

After this debunking, the director of the National Széchényi Library, Gábor Mátray, purchased most of the controversial documents in an attempt to overcome the confusion around them and placed them in two wooden boxes in the library's Manuscript Collection, where they still reside today. In this way, he withdrew them from circulation and preserved them for proper philological analysis.[7]

When I first viewed these wooden boxes, they had barely been touched for 150 years. When they were brought out to me, their contents revealed an extraordinary collection of one- or two-page letters and small booklets. I found family and royal documents, maps, "eleventh-century" prayers in Hungarian and Latin, a "Tartar" letter, a piece of "pagan writing" that, according to the inscription, had been found in a "cave" in 1702, the Banská Štiavnica codex, which contains prayers and notes on local gold and silver production, a twelve-page booklet "from the 17th century" on the members

and Chinese languages were from the same root because the Hungarians called the Chinese *csinok* and considered them good-looking, which is where the Hungarian word *csinos* (pretty) comes from.

> I often have ideas that I have never heard from anyone, nor have I read, and yet I believe in them, and there is an example for that: when our ancestors were closer to China, and thus witnessed the habits of the Chinese, in these times our Scythian Fathers called the Chinese people *Csinok*. Among our fathers, who were peasants and shepherds, only a few could wear beautiful colored Chinese clothes, and who could afford to wear such clothes, as the Chinese wear even today, so they were called by our ancestors *csinos* [pronounced "chinosh," meaning pretty], "Chinese-like."
>
> And the Chinese are very mischievous, they throw stones at strangers, they are deceitful, they steal things, and they could not have been much better in the past either. Our simple-minded ancestors who, leaving their natural innocence, behaved badly, were called *csin* or *csinos*, i.e., full of the sins of the Chinese. These sins were not known to our ancestors, and we later forgot the meaning of this word, and today use it in its opposite meaning. Now bad children are called *csintalan* (without *csin*), that is, lacking fault, especially in Transylvania.[10]

Though linguistics was significantly less developed in the early nineteenth century than it is today, the idea that two words from two different languages could be related solely based on their acoustic similarity was considered unfounded even at that time. Against this background, it is easy to imagine that Literáti—being proud of his self-attained, though scant, knowledge—might have failed to identify the documents he came across as forgeries.

The pseudo-antiquities attributed to Literáti have two common characteristics. First, they are usually relatively short: the Banská Štiavnica codex, the longest of them, is only eighty-four small pages. Most of the other counterfeits contain no more than two pages of text. Whoever made them could not have spent more than a few days on each document, and he (or she?) presumably preferred creating portraits, coats of arms, and maps to writing texts. Second, each item in the collection claims to be something old, rare, and precious, usually a Hungarian linguistic monument.

As I was putting these documents back into their box, a rising doubt came over me. It seemed improbable that the Rohonc Codex could really belong to this group of forgeries. I read all five volumes of Literáti's diary together with his letters to his commissioners and his bills, becoming very familiar with his sloppy handwriting in the process. He copied many special characters into his diary and described many mysterious writings in his letters—the books he had laid his hands on and the ones he had forged left their marks in these notes—but nowhere did he mention the distinctive writing of the Rohonc Codex. In contrast, the piece of birch bark carrying "Hunnish writing" that Literáti mentions in his diary can be identified as the "wooden book from Túróc." While most forged documents in the Széchényi collection were proved to have originated in Literáti's workshop or to have been connected to him in some other way, I found no evidence to suggest that Literáti had ever even heard of the Rohonc Codex.[11]

In his long 1866 article, Károly Szabó claims that he does not believe the Rohonc Codex to be Hunnic and argues that it is likely to be the work of "Sámuel Literáti Nemes, who is known to be a skilled forger."[12] Just how likely (or unlikely) this is can be determined by examining Literáti's other activities and undertaking a comparative chronology of the codex.

Literáti's career reached a belated climax around 1828, when he moved to Buda. This is when he finally went so far as to offer his services to members of the aristocracy. He didn't even open his shop until 1835. Theoretically, it is possible, of course, that Literáti passed the codex on to the Batthyány library quite unobtrusively, and that it was transferred to the manuscript collection of the Academy not quite ten years later, surrounded by great excitement, but with no one around with any idea how it was given to the Batthyánys in the first place.[13] This is possible but unlikely—so much so that further evidence would be needed before we could entertain this scenario seriously.

It is worth asking at this stage why we regard the codex as a forgery. It possessed neither of the two traits noted above—brevity and rarity—that characterize the contents of the wooden boxes. Most of the forged documents in those boxes were only a few pages long and could easily have been created within the span of two days. The 448 fully scribed pages of the Rohonc Codex required a great deal more time and energy to create. And it does not pay to create extensive forgeries, since these are known to be easier to expose thanks to the number of repetitions in the text. This is precisely why the Phaistos Disk, with its limited number of characters, is unlikely ever to be deciphered. We must also ask ourselves whether we have any good reason to question the authenticity of the codex in the first place. A document is considered counterfeit if it tries to pass itself off as something other than what it really is. Thus a one-page piece of gibberish forged by Literáti aspires to look Chinese. Another pretends to be a prayer from the period of King Andrew I, when it is in fact some eight hundred years younger. A third, which was colored by Literáti or one of his contemporaries, wants to pass as a collection of miniatures from the fourteenth century. The Rohonc Codex, by contrast, does not wish to look like anything but what it really is: a manuscript with a series of strange symbols. It is true that once it emerged, many believed it to be an ancient Hungarian or Hunnic script, but this actually says more about that period of Hungarian history than about the codex itself. The mid-nineteenth century saw the emergence of a succession of historical sources, authentic or not. The codex itself, in any case, never claimed to be an ancient source of historic importance. Actually, it does not claim to be anything at all. There is only one thing that we can take for granted, and that is that the illustrations on its pages indicate a religious topic.

We cannot even be sure that it wished to appear older than it was. It may have been written in the sixteenth or seventeenth century, something that we suspect because of the sixteenth-century watermark on its paper. But if it is indeed from the nineteenth century, we still cannot say that it was trying to deceive readers about its age.

And what makes us think that the codex either is written in one of the natural languages or else is a meaningless text pretending to be a natural language? Nothing at all indicates that its author wanted it to look like a *script*, whether in ancient Hungarian or some other language. There are no clear word boundaries, the symbols are not typically letterlike, and there are too many different symbols anyway: more than 150 frequent ones, not counting those that appear only once or twice. What this system of symbols does resemble is a cipher, a code, or an artificial-language scheme popular in the early modern era.

THE HOAX THEORY

Whether the codex is authentic is a crucial question with great implications. If we conclude that a script is merely a hoax, then no time should be wasted on trying to decipher it. In such cases, however, the question of *cui prodest* (who benefits?) cannot be avoided. Whose interest is served by creating a pseudohistorical source? What could have been the intention of the author? Was he simply trying to make money? To ridicule someone? Was he driven by malice? Or was he a lunatic who felt compelled to transmit the message of aliens from outer space?

The hoax theory emerges from time to time in connection with all ciphers or code systems that refuse to be solved—and sometimes even with codes that are eventually successfully broken.[14] The Beale ciphers are a well-known example that provoked heated disputes. These writings were found in the 1880s and they allegedly describe the location of buried gold treasure that had been hidden decades earlier. The Beale ciphers are actually a set of three code strings consisting of numbers. The second text was already decrypted at the time of its publication: it uses the American Declaration of Independence as its basis, the words of which are numbered in the order in which they appear therein. Each letter of the plaintext is then substituted for the number of the word in the Declaration that starts with that letter. This is what is known as a *book cipher*, one that is nearly impossible to decipher because of its structure, unless we happen to know which edition of which book is being used as the basis for the code. It should come as no surprise, then, that the first and third ciphers have not yet been broken, despite the best efforts of numerous gold diggers and codebreakers, adventurers and mathematicians. No one has been able to identify the book that would help crack the code. Or if they have, they have quietly and happily dug up the treasure without publicizing their success.

Many have observed, however, some contradictions in the romantic background of this treasure hunt. To start with, why does the second (already deciphered) cipher announce that the first cipher will reveal the location of the treasure, and that the third will list those who should receive their share of it? How did the author know in advance that the second text would be deciphered first? What was the point of enciphering the second text if the location of the treasure is revealed in the first? And why does the second cipher refer to the other two as first and third? Moreover, how could the third cipher, supposedly containing the names and addresses of thirty of the late treasure hunter's next of kin, contain only 618 characters? Such a small number of characters would not be enough to encipher the sixty names, not to mention the addresses. Taken together, these peculiarities strongly suggest that the Beale ciphers are a hoax.[15]

Still, how can a series of numbers ever be proved to be a fraud, something that can never be decoded? Codebreakers are unanimous in their answer to this question: it is theoretically impossible to prove that a cipher or a code has no meaning and will never be solved. Curiously enough, the same position is taken by those on the other side of the debate, those who believe the Beale ciphers are not a fraud or hoax. When we assign the numbers of the first coded Beale text to the words of the Declaration of Independence, the result is unintelligible—except for one specific section, where the letters appear in neat alphabetical order. This is so improbable by the laws of statistics that it could only have been done intentionally. Some conclude that this pattern is a sign that the fraudster grew tired and started following an ordered string instead of a random one.[16] Others, by contrast, suspect that this is a hint in a *doubly* enciphered script, meant to encourage the codebreaker to continue his work.[17]

The Voynich manuscript has also been suspected of being a hoax, though it has received no shortage of proposed solutions. Imaginative codebreakers have suggested a wide array of possible authors, including Roger Bacon, the Cathars, the Khazars, the Ukrainians, the Jews, and even God himself, and they have proposed different topics, among them a nebula, botanical novelties of America, contraception, and suicide, with varying degrees of correlation to the odd pictures that abundantly illustrate the manuscript. Perhaps it is due to the confusing mass of theories, or perhaps to the succession of failed attempts to break the manuscript's code, that those suspecting fraud have made their case. It is highly unlikely that a fifteenth-century author, so the skeptics argue, would have been able to devise a cipher that could withstand the efforts of the best codebreakers in an age when historical ciphers are typically broken in a matter of days.

So who could be the person behind this hoax? There are many candidates, ranging from the sixteenth-century medium and alchemist Edward Kelley to the twentieth-century book collector Wilfrid Voynich himself. In most cases, the candidate's motive is clear: financial gain. An intriguing and mysterious book could be sold for a fortune. Kelley was financially supported by the gullible English mathematician and magician John Dee and by the generosity of Emperor Rudolf II. Voynich the bookdealer earned a living by selling interesting books to enthusiastic customers.

It is easy to claim that such a document is a hoax. Proving that claim is much more difficult, especially when we consider the peculiar language of the Voynich document. When the billowy characters are transcribed into Latin characters according to the usual equivalence system used by Voynich experts, an average paragraph reads something like this:

tshedor shedy qopchedy qokedydy qokoloky
qokeedy qokedy shedy tchedy otarol kedy dam
qckhedy cheky dol chedy qokedy qokain olkedy
yteedy qotal dol shedy qokedar chcthey otordoror
qokal otedy qokedy qokedy dal qokedy qokedy skam[18]

The number of repetitions here makes us wonder if what we are looking at is really a natural language or a coded text. Natural languages just never look like this. Deeper analysis, however, does prove that the Voynich language contains patterns that are not much easier to create than a real code. Some combinations of letters are more frequent, while others never occur, and some characters tend to appear at certain parts of a line, at the beginning or the end. Words—or the combinations of six to eight characters that appear to be words—seem to obey their own morphological rules. Lines have their own structures, too; words that fall at the end of a line, for example, are usually shorter than words at the beginning or middle. A series of statistical examinations and comparative linguistic studies indicate that this language is not a random string of characters, nor does it display the characteristics of poems, highly repetitive magical chants, or the pathological speech structures of the mentally ill.[19]

Some researchers who insist on the hoax hypothesis, however, have designed methods that require only a table and a grid that can be superimposed on the table—tools that were accessible in the sixteenth century. Supposedly, these tools can be used to create a meaningless yet structured text, similar to the Voynich language.[20] The jury is still out, but the hoax theory definitely remains a strong contender.

So what do all these other ciphers mean in relation to the Rohonc Codex? The botanical, astronomical, and "bathing ladies" illustrations of the Voynich manuscript refer to some mysterious knowledge. The manuscript as a whole is attractive and almost certainly could have been the pride of any book collector, though it eventually landed in a public collection (the Beinecke Library at Yale University) without making anyone rich. The Rohonc Codex, by contrast, appears to be a dull book of little value.

Its images indicate a religious subject, not very fascinating in a time when prayer books were sold on every corner.

Of course, the codex would have been more valuable had the images pointed to some exciting New Testament Apocrypha. Such texts were quite popular even in the nineteenth century, before becoming the subject of tremendous, outsized interest on the part of twentieth-century scholars and adventurers. But the images in the Rohonc Codex, as we shall see, suggest no such thing.

Having discarded a financial motive, we can still speculate that "some prank-making mood was added to romantic zeal," as Kelecsényi puts it in connection with the little-known "Gelle prayer book" from northern Hungary. A parish priest in a small upper Hungarian town mentions in a 1777 letter a special prayer book written in unknown characters and owned by an old man. The priest allegedly copied the book carefully before the old man and his book were both lost on a pilgrimage. The priest titled his copy of the book "A booklet of everyday prayers, property of a decent old man; here one can find the creed of Saint Athanasius written with Scythian-Hungarian letters, moreover, the Lord's Prayer, the Hail Mary, the Creed, the Ten Commandments and the Confiteor, together with the psalms of repentance; all of these were written in the Jewish way from right to left, using an alphabet that is to be found on the next page" (fig. 2). Events followed the usual scenario: in the first half of the nineteenth century, a number of literary historians believed in the authenticity of this copy until the philologist Károly Szabó came along and disappointed them by using scholarly evidence to reveal this "relic" as a fabrication of inferior quality. What Szabó discovered is that the "Scythian" text was actually the transcript of a printed and at that time available book (János Kájoni's 1676 *Cantionale catholicum*) in Hungarian runes.[21] On the basis of this discovery, Szabó argued that the "ancient" prayer book could not be very ancient at all, and that "some priests must have wanted to play a prank on István Hájos, teacher at the Piarist school of Kecskemét, by making up a fantastic background story and creating these fake characters." Hájos duly went on to write an analysis on the prayer book that was never published. Its Latin title can be translated "On the Scythian-Hungarian-Sekler Language."[22] I had a hard time locating a copy of this study. Several librarians helped me find my way through a series of inaccurate nineteenth-century references until I finally discovered it in the middle of a thick book in the Library of the Hungarian Academy of Sciences.[23]

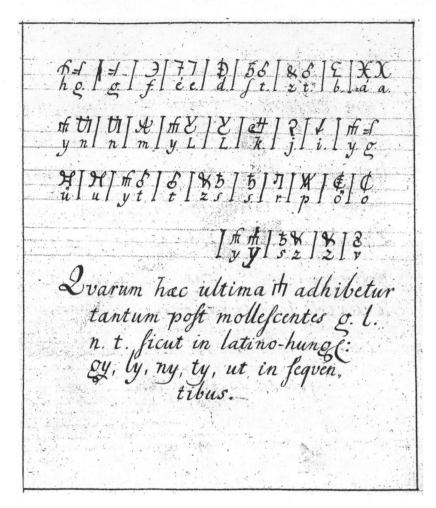

FIG. 2 | The table of characters of the Gelle prayer book. Library of the Hungarian Academy of Sciences, MS Történl. 4-r, no. 38, p. 135.

The similarities between the Rohonc Codex and the Gelle prayer book are striking. Both contain prayers and religious texts, both read from right to left, and both contain mysterious characters. Are both the product of a practical joke? Before jumping to conclusions based on the few similarities between the two texts, we should enumerate the differences as well. First of all, the Gelle prayer book tries to look like an important literary relic. It uses, by and large, an alphabet that was known at the time, a Hungarian runic script composed of fewer than thirty characters (fig. 3). The Rohonc

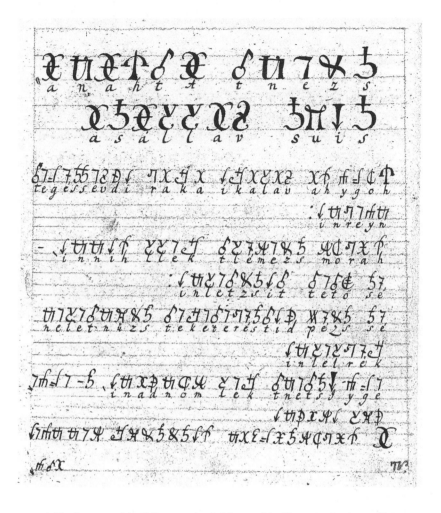

FIG. 3 | The first page of the Gelle prayer book. Library of the Hungarian Academy of Sciences, MS Történl. 4-r, no. 38, p. 132.

Codex, on the other hand, appears to be mystical but doesn't wish to pass as an important historic literary text of Hungarian culture (or any other culture, for that matter). Second, while the Gelle prayer book is basically a monoalphabetic text, which uses the characters of one alphabet of limited size and could have been deciphered using a simple statistical analysis, the Rohonc Codex is built on a complex system of symbols of several hundred characters, looking not at all like an ancient natural language, and it does not yield to simple deciphering methods, as we shall see in the next chapter.

It seems that the Codex of Rohonc is not a simple forgery; nor is it a practical joke. The problem with the hoax hypothesis is that the patterns of the character strings show too much regularity and consistency. It is possible to imagine a fanatical author who would write nearly 450 pages of gibberish, even when a shorter text would suffice. It is, however, much less plausible that someone would go to such great lengths to concoct a meaningless text displaying such consistent patterns. And consistent patterns there are, appearing many pages apart in the text, and even in those sections that appear to have been written in a different person's handwriting. Perhaps *not one but several* fanatical authors together agreed to follow certain "linguistic" rules and to copy meaningless symbols according to fairly strict guidelines? Although the possibility that the codex is a hoax can be ruled out definitively only when the solution is found, already at this point we can claim with confidence that it doesn't fit either into Literáti's profile of forged historical documents or into the class of historical bluffs.

I will return to the hoax theory in chapter 8, where I examine invented languages described in imaginary travel reports and in testimonies on faraway, nonexistent cultures. Apart from that one chapter, however, I will assume that the language of the Rohonc Codex consists of a meaningful, nonarbitrary code, an artificial language that can actually be decoded.

CHAPTER 2

CODEBREAKING ATTEMPTS

BEGINNINGS

A certain ill fate looms over past attempts to decipher the Rohonc Codex, despite the surprisingly promising first endeavors. The Hungarian language historian János Jerney first examined the codex in the early 1840s, shortly after it was found. Judging by the watermark, he soon identified the paper as coming from sixteenth-century Italy. This type of paper was indeed fairly common in early modern Hungary. From the biblical nature of the illustrations, Jerney concluded that the author must have belonged to a Christian culture. He then went on to compare the writing to various Asian writings, since it seemed to display some Eastern characteristics. Jerney was not yet tempted to regard the source as an ancient Hungarian document. Instead, he suspected that if the language was a natural one, it could have been written by Tartars, who had settled in medieval Hungary and become Christians. His theory was that the Tartars used their own Asian letters to write the Rohonc Codex or the original book from which the codex was copied. Jerney did not think the purpose of the book had been to "deceive coming generations or to create a counterfeit just for the sake of a game," but he did toy with the idea that the text might be a cipher.[1]

In the following years, a succession of scholars, Hungarians and foreigners alike, tried to identify the symbols: Ferenc Toldy and Pál Hunfalvy from Buda, Josef Jireček from Prague, Bernath Jülg from Innsbruck,[2] Alois Müller from Graz.[3] Later, Mihály Munkácsy, the famous Hungarian painter, even took the codex with him to Paris to have it examined. The first systematic and published attempt to break the code was by Kálmán Némäti, who started working on the book after it was brought back from its eighteenth-month sojourn in Paris.

Kálmán Némäti (1855–1920), the "educator of the nation" (as he called himself), had a life so unique that it should be described in a separate monograph. He certainly did not belong to the institutional mainstream of Hungarian literature or historiography. After he gave up on educating the nation, Némäti spent two years in an empty bear cave, where, according to his entry in a biographical encyclopedia, "he wore underclothes and a monk's habit made by his own hands; ate wild fruit and roots; and was often visited by the people of the land, who would listen to his speech and give him wheat, fruit, and bread."[4] Later, living on alms from his relatives like a "beggar-writer," he published a long series of articles not only on the Khazars, the Turks, and the origins of the Hungarian people but also on a proposed reform for teaching the alphabet in primary schools and the laws of nutrition (he himself was a hardcore vegetarian). As for the Rohonc Codex, he correctly identified the writing as running from right to left, and incorrectly argued for the ancient Hungarian origin of the text. He published his views on his own. In addition, he submitted a handwritten classification to the Hungarian Academy of Sciences that listed and grouped the symbols of the codex, of which he had found almost eight hundred. This large number of symbols made him suspect that they stand for syllables rather than letters.[5]

Némäti's research received some scholarly attention when he requested a grant from the Academy. Its Committee on Hungarian Linguistics took his proposal seriously, and, according to the official record of their meeting on 12 November 1898, the committee members decided that they primarily needed a "paleographic study in order to judge whether the manuscript is an ancient Hungarian source." So they asked four paleographers, experts on ancient writings, to evaluate the codex. These experts examined a variety of evidence and concluded that although the paper "was indeed from the first quarter of the 16th century, the writing on it is a later forgery." Their main argument was that "it is impossible to encipher a text using 900 symbols because no man on Earth could possibly read such a text, not even the person who had created it. Handling an alphabet of 900 secret symbols is beyond the capacity of human memory. The words are not separated, making the text difficult, even impossible, to read. Moreover, no corrections have been made, which is unheard of in a manuscript of this length." Two of these observations are simply incorrect: there are ciphers that use nine hundred symbols or more, and the codex does contain a number of corrections and deletions. Still, the paleographers clearly did not believe that Némäti's ancient Hungarian script hypothesis could possibly be correct. Instead, they examined whether the string of symbols could be a cipher. The committee's conclusion was unequivocal: "All of these convinced the committee that Mr. Kálmán Némäti has wasted his ~~rare~~ tremendous zeal on an impossible task, and that anyone encouraging him to continue this work would do him a disservice" (struck-through word in the original record).[6] With this declaration that the Rohonc Codex was a forgery, the Committee on Hungarian Linguistics of the Hungarian Academy of Sciences put an end to a long series of attempts that had in any case already lost their initial fervor and were becoming more doubtful. The Academy's pronouncement discouraged would-be Rohonc codebreakers for almost a century (with a single exception). Upon the arrival of the third millennium, however, many voices broke the silence.

THE ANCIENT HUNGARIAN THEORY

Attila Nyíri, one of those voices, is neither a professional historian nor a paleographer but an electrical engineer. This in itself would not be a problem, because important insights into the breaking of historical coded texts often come from nonspecialists. In the late 1990s, when copies of the codex were not yet widely available, Nyíri used the two complete pages to which he had access. He read the symbols on those pages as a prayer written in ancient Hungarian letters—that is, a natural language. In other words, he did not interpret the characters of the codex as *corresponding* to the letters of the Hungarian alphabet; he simply *recognized* them as letters of a natural language that could be read spontaneously.[7] He happened to read the text upside down, something that he himself realized later, but it was not this mishap that proved him to be completely wrong. For one thing, Nyíri allowed one letter (one sound) to correspond to several symbols, a perfectly common method in ciphers, though not so much in natural languages. For another, he read the same character as several different letters. He furthermore claimed that the order of the letters was sometimes jumbled up in the text. This method of decoding, however, is so arbitrary that it makes any deciphering attempt successful by definition and thus allows any text to be read in any way.

Turning the page upside down, a few lines from the Rohonc Codex as Attila Nyíri read it, from right to left in "the ancient Hungarian language," sound like this:

> Your God has arrived. Oh, the Lord is flying. There are the holy
>   angels.
> Oh, yes, them. Sung with decorous words, send the song, pour it.
> The Lord is to come, I am flowing everywhere. Oh. The Lord is
>   honored.
> The peace of the Lord flies far. Those holy words . . .

THE DACO-ROMANIAN HYPOTHESIS

Nyíri was no philologist, and he had access to only a small section of the original text. These qualifications cannot be made in defense of Viorica Enăchiuc, a Romanian archaeologist who summarized her twenty years of studying the Rohonc Codex in an eight-hundred-page book.[8] Enăchiuc claims that the text is written in Vulgar Latin—a language that she also calls proto-Romanian—from the eleventh century. She transcribed the complete string of symbols, created a dictionary of this hitherto unknown language, published a complete reproduction of the codex (without ever asking the library for permission to do so),[9] and then translated the text into modern Romanian and French. Furthermore, she made room in her voluminous book for studies in Romanian (and their French translations) on topics such as the "musical notes" and "musical content" of the codex, various maps of Dacia, and even the author's list of publications.

Only one thing was left out of this thick volume: the basis of all successful solutions, namely, the one- or two-page key specifying which symbols of the codex correspond to which letters of Enăchiuc's proto-Romanian alphabet. Enăchiuc hints that this correspondence table will appear in a "second volume" to come. If readers attempt to create such a key on their own, however, they will quickly discover that each symbol in the codex corresponds to a different letter every time it appears, carefully tailored to the meaning that Enăchiuc attributes to her "Vulgar Latin" text.

Throughout the twenty years she spent studying the codex, the archaeologist never realized that she was reading the text the wrong way. She did notice that the symbols read from right to left, which is obvious from the fact that the text is aligned with the right-hand margin and that hyphens sometimes fall at the left end of the lines. Why she read the text from the bottom up, however, is hard to explain. Even a quick study of the text identifies coherent strings of characters that, when broken at the end of a line, always continue at the right-hand end of the line below. Furthermore, there is the fairly conspicuous fact that the end of a chapter or section is followed by a blank area at the bottom of a page, not at the top. Given these obvious features of the codex, even the earliest researchers could correctly determine the direction of the writing, right to left, top to bottom.

Equally bizarre is the fact that Enăchiuc failed to notice that certain symbols always stand together, such as the frequent IO:O and several dozen other examples that she decodes as separate letters every time they occur. If they really were a string of separate letters rather than a complex character carrying a certain meaning, then they would be signs of a level of structure so high as to be unmatched by any other language. In other words, the number of letter combinations that always stand together, like $q + u$ in Latin, is much too high. Enăchiuc also believes that the second symbol of one specific digraph is a sentence delimiter. But if that were the case, then all sentences would start with the same letter, the first symbol of this digraph (remember that the text reads right to left). This weird feature of Vulgar Latin went unnoticed during the process of translation only because the same symbol is always translated by different letters. Thus Enăchiuc provides yet another method that enables us to translate any kind of ciphertext in any way we choose.

All of these peculiarities make sense once we read Enăchiuc's reconstruction of the text and discover her motive. In her rendering, the codex describes the centralized Blaki (early Romanian) state, located between the Tisza and the Dniester Rivers and led by the emperor Vlad, at the peak of its glory in the eleventh and twelfth centuries. The codex contains speeches, prayers, and songs in connection with this state, especially battle songs meant to inspire the eleventh-century Blaki youth to glorious victory over the Oghuz and Hungarian people. The Oghuz, if I understand Enăchiuc correctly, are in fact the Pechenegs, who were allies of the Hungarians and posed a threat to the centralized Blaki state as well as to the Byzantine Empire around the year 1100.

Let us read the transcript of the codex by the Romanian archaeologist Viorica Enăchiuc. It reads, from right to left and bottom to top in Vulgar Latin—that is, in the Daco-Romanian language—"DETETI LIS VIVIT NEG-LIVLU ITI ITI ITIA NITEREN / TITIUS SUONARES IMI URAST UCEN."[10] Chances are that the reader would not quite be able to read this in Vulgar Latin, a language that, to the best of my knowledge, has no other surviving source, let alone a grammar book or dictionary. This does not bother Enăchiuc in the slightest, as she has created a detailed dictionary in which she assigns each word of this language to another word, usually Latin. For example, the Vulgar Latin ITI comes from the Latin *eo*, *ire*, *ivi*, *itum*, which means *go*. The strange-looking NEGLIVLU is not listed in the dictionary but could be the verb *neglect*, based on a similar-sounding word, the Latin *negligo*. SUONARES stands for "the Hungarians" because it sounds like Hunor, a character in *Gesta Hungarorum*, an early history of the Hungarians written by a medieval Hungarian chronicler. Enăchiuc translated the Vulgar Latin lines above into French, and you, dear reader, can read them in English: "In great numbers in the fierce battle, go without fear, go heroically! / Advance thunderingly, to sweep away the Hungarians and win!"[11] Elsewhere, Enăchiuc quotes a motivational speech that was delivered at the fort at Inau, Transylvania, before the battle over the River Tisza against the Hungarians:

A SUOAR NOAS SUOAR STRIOL

INOU IU OI IURA FIDI TENIS NITIOI

INOU NEVI TENES SEDANI DIT =

IU ELICEN VASI ABDI BINI

SUNAR EDO LIDI SUNAR TITI TISA

TI INOU TO VEIKI UTI NITITI ACIRA

TI DETI ATR DIRA SATI SUNARA

OT NIS TENEN VI ULCER IURAI SUNAR

DICA ER ÚTI VEIK IUKU INOU A ROI

SUOAR OSORAI SUOAR STRIOOL
ISTI IS ETIA VI IKER UTI ITI SER[12]

In English, this reads:

> In our defense, in defense of the Strei! Go to Inau and swear!
> Defend it with glory and defend the united Ineu
> in continuum, completely.
> Go together, I have pushed forward. Together fight back
> the Hungarians; I encourage you to fight over the Hungarians, not
>   letting go of the Tisza at your Ineu; push forward, to shine with
>   glory, by your bravado
> stop the cruel tragedy caused by the Hungarians.
> To defend us strongly, swear to wound the Hungarian.
> Decide my lord, to push forward at Ineu with a hawk's cry again
>   in defense,
> decide about the defense of Strei in advance.
> Go and now you will strike with greater force, now that you go
>   united.[13]

A few remarks: Strei, which should be defended, according to this text, is the river known in Hungarian as the Sztrigy, which happens to run several hundred kilometers from either Ineu (Borosjenő), or Inau (Ünőmező), not to mention the River Tisza. I also wish to note here that the song sounds no less gritty in French translation.

Enăchiuc's reconstruction of the centralized eleventh-century Romanian state and its soldiers who would defeat the Hungarians over and over raises a number of historical problems, and it is little wonder that even Romanian historians have criticized it.[14] Let us for a moment accept this theory and assume that ever since Roman times there has been a Romanian state, and that its people spoke "Vulgar Latin." It is still impossible to accept Enăchiuc's rendering of the codex: she reads it the wrong way, she decodes the same strings of symbols in different ways in different sentences, and she makes up a nonexistent language that has sources nowhere else.

Although the illustrations in the codex will be discussed in detail later on, we must now address the peculiar fact that Enăchiuc sees the emperor Vlad, his subjects, and the ambassadors from Byzantium arriving for an audience with the emperor in images that every other researcher identifies as typical biblical scenes. There is an image, for example, of Christ entering Jerusalem on a donkey, with people laying their clothes on the road before him, a palm tree from which people have removed the leaves, and also the money changers being driven out of the Temple (fol. 14v). Enăchiuc gives the caption for this drawing as "Vlad preparing for the alliance to be made with the Byzantines against the 1064–65 conquest of the Oghuz; the metropolitan archbishop of the Blak, Sova Trasiu, blesses the warriors in the temple with battle signs." Another image, depicting Mary, a winged angel, and Joseph (fol. 16r), bears the caption "Sova Trasiu, metropolitan archbishop, in a wooden church with a bell tower, sending a book to Jaroslav I, Prince of Kiev, so they would unite with the Blak in the war against the Oghuz." On the right, where I see Joseph and the angel, she sees "the Prince of Kiev, who, receiving the news, accepts the alliance." The Adoration of the Magi, which, to be clear, also depicts the Star of Bethlehem (fol. 21r), Enăchiuc sees as "Vlad, head of the Blak, standing with Sova Trasiu, metropolitan archbishop, and a general, receiving the envoy of Byzantine emperor Constantine, in an army base somewhere in the Sub-Carpathian region."

She does not stray so far from the traditional rendering in the obvious scene of the Crucifixion (fol. 26r), where she provides this caption: "Vlad, governor of the Blak, wearing a helmet and weaponry, is praying for victory at the foot of the crucified Savior, before leaving for the war against the Oghuz." But the arrest of Christ in the Garden of Gethsemane (fol. 28v), according to Enăchiuc, is actually the meeting between the Goths, who were crossing the Blak territories, and the Byzantine army leaders. And instead of Christ meeting Pilate (fol. 40v), Enăchiuc sees Vlad before the Byzantine emperor Alexander I. The scene depicting Christ wearing a crown of thorns standing before Herod (fol. 45r) is explained as the Cuman king receiving the blessing of the Cuman high priest. The king is about to leave for battle, where he is going to fight on the side of the Blak and the Byzantines. The events unfolding around Christ's tomb (the *noli me tangere* scene, the empty cave with the angel, fol. 54v) Enăchiuc interprets as the scene in which the Goths are preparing to fight the invasion of the Pechenegs. All of these illustrations are typical and unambiguous depictions of biblical scenes using conventional iconographic symbols. The appendix of this book shows the list of these scenes in the codex, while chapter 4 contains an analysis of them.

THE SANSKRIT THEORY

We know that our perceptions are deeply influenced by a priori nationalistic expectations (among other factors), a fact illustrated not only by Enăchiuc's bizarre reading of the Rohonc Codex. In 2004, *Turán*, a journal specializing in research on the early history of Hungary, published a solution by the Indian researcher Dr. Mahesh Kumar Singh. Dr. Singh claims to be a descendant of the Hunnish royal family. The background story of his codebreaking solution is telling. The comments by *Turán*'s editor-in-chief are worth quoting at length.

> The last few months have been most trying for the *Turán* journal, which was in its eighth year. Perhaps it has also launched its most fertile period. We had just decided, in October 2004, to navigate more toward natural Hungarian culture, when a hitherto unduly neglected area of research opened up for us. . . . The conversation went on into the night and as a result a small window was opened onto the many millions of people with Hunnish and Scythian origins, who, though they lived in India, were still keeping part of their traditions. . . . During the meeting that was held in the Frontvonal coffee shop in Budapest, we agreed with Rozália Hummel and Dr. Mahesh Kumar Singh that we would travel to Rajastan and Gujarat in early March to pave the way for a research expedition that was planned for the autumn. As we were saying goodbye, my phone rang. It was a dear friend from Aszód. "Would you like to see a nativity play that was to be performed by the members of a pipe orchestra from the Galgamente region?" he asked. We had mangalica salami in the pantry and homemade palinka to go with it—we were ready for a party. We invited Dr. Mahesh along too, since he had never attended such an event. We hadn't either, we soon found out! It was fantastic! We saw an authentic piece of our folk tradition, a shepherd's nativity play. There we had eight true Hungarian men, in wide trousers, boots, braided hair, goat's pipe, and a cow's bell. This was the performance of recent years for me, and I hope it will be for many more people in December 2005 too.
>
> The nativity folk left, but they left with us a piece of our common past, present, and future. Then I got out a book that was published

in the *Turán* book series, Imre Harangozó's *I Have Been Thinking a Lot About Our Previous Fathers*, along with a couple other
earlier issues of *Turán*. I showed to my Hunnic friend the folk cos-
tumes one could still see on shepherds, cattlemen, and horsemen
in Hungary in 1995—and maybe even today. Dr. Mahesh Kumar
Singh, browsing through the fourth issue of year seven, started to
read in English without a pause. I asked my wife, Gyöngyike, "Has
Mahesh learned Hungarian?" She then showed me the *Turán* as it
lay there, open, with the facsimile of the still undeciphered mysterious Rohonc Codex in it. Mahesh was reading it, just reading
nonstop. I stood there, perplexed, my heart beating faster; I had all
kinds of thoughts; "God is with us, He is showing us the right path,
we need to believe and to fight, we should never give up hope. This
is a great discovery and it may direct the attention of the international academic world to research on early Hungarian history, on
our Eastern roots."[15]

The editor of *Turán*, despite contrary advice from some fellow editors,
published the first twenty-four pages of Singh's transcript, along with a Hungarian translation.[16]

Here is a sample of Singh's version of the codex, which he reads from left
to right, top to bottom, in Sanskrit, which the editors of the journal translated into Hungarian (which I have translated into English):

Oh, Lord, the people here are very poor, sick and hungry,
give them skills and strength to satisfy their needs
provider, do not harden your heart
do not take your hand from this needy people
their needs that they desire for themselves
grant these to them for their sake
whenever you help them these people
that they may find this help perfect.

We must give credit to the editors of *Turán* for publishing a detailed
refutation from a reader in a subsequent issue. This reader pointed out
(much as I and others have done with regard to Nyíri and Enăchiuc) that
Singh decoded the same strings of symbols in different ways: for example,

he interpreted one digraph in eleven different ways, and gave one other, longer string of symbols four separate meanings in four different places in the text.[17]

SYSTEMATIC ATTEMPTS

We are not quite finished summarizing earlier attempts at decoding the Rohonc Codex. Although none of them was undertaken by a professional codebreaker, they are relevant because of the tools and methods used, methods that would also be used by professionals. And although they were considerably more successful than the attempts described so far, they have modestly refrained from claiming to be a full solution to the puzzle. The first such attempt is that of Lieutenant Colonel Ottó Gyürk, who became known in 1969 for his success in deciphering the numbers in the encrypted diary of the novelist Géza Gárdonyi. Gyürk was commissioned to work with Gábor Gilicze, then a university student, who had deciphered the text of Gárdonyi's diary.[18] Together, they successfully transcribed the secret text, after which Gyürk felt confident enough to have a go at the Rohonc Codex.[19] Studying the continuous strings of symbols that break at the end of lines, as well as the incomplete lines and pages, he quickly realized that the text reads from right to left and top to bottom. He also correctly identified as a hyphen the two parallel lines that often appear at the end of lines on the left side of the page. This symbol (which looks like an equals sign in English) is familiar as a hyphen to researchers studying handwritten texts from the sixteenth to nineteenth centuries, but Gyürk was not a philologist. Following the procedure that proved to be successful with Gárdonyi's diary, Gyürk tried to identify certain strings of symbols as numbers. He confessed to spending years creating hundreds of systematic statistics, yet he was unable to crack the code. Still, Gyürk's method, which focused on the patterns and statistical features of (groups of) symbols, was the first approach to deciphering the Rohonc Codex that held promise.

A similar, but computerized, analysis was carried out by Miklós Locsmándi, who published his results in the same 2004 issue of *Turán* that also contained Singh's transcript. Locsmándi apparently received less support from the editor. He realized fairly quickly that the language of the codex is a constructed rather than a natural language, and he was willing to let go of the early Hungarian script theory. He distinguished between the simple and the complex symbols in the codex, a real achievement in itself; neither Némäti, nor Nyíri, nor Enăchiuc had recognized that certain symbols or characters always stand together and must be translated as one unit. Locsmándi then examined the frequency of the symbols in order to determine whether they represented letters, syllables, or perhaps complete words. He drew attention to a queer feature of the codex, namely, that the text contains repetitions that are too frequent to occur in a natural language or an enciphered text. He concluded that "the language of the text probably uses a small number of simple basic units" and that this structure is strongly characteristic of "prayers of litany, perhaps charms from the folk tradition," which suits the proposition, based on the illustrations, that the codex is the prayer book of a religious or sectarian movement.[20] But Locsmándi was also unable to crack the Rohonc code.

This is where things stood when I started my investigations around 2006. Little did I know that two young Hungarian researchers, Gábor Tokai and Levente Zoltán Király, were already working on the Rohonc Codex, and that they would eventually arrive at a collaboration and, ultimately, a real breakthrough.

CHAPTER 3

THE QUEST BEGINS

THE CODEX AS AN OBJECT

Little is known about the history of the Rohonc Codex as an object, despite the fact that the past 150 years have yielded some promising information. One can, of course, try dating the paper on which it is written using the standard dating tool of watermark identification. Previous researchers combined the partial watermarks that can be found on the pages of the codex and identified the resulting image as one of the watermarks in Charles-Moïse Briquet's massive four-volume catalog of watermarks (1907). They concluded that the codex was made between 1530 and 1540 in northern Italy, more precisely in Vicenza or Udine.

When I looked more deeply into the problem of watermarks, however, I could not confirm this conclusion beyond any doubt. When you hold the leaves of the codex up to the light (e.g., fols. 11, 12, 13, 24, or 27), you can see what earlier studies of the codex identified: a picture of a star and an anchor. According to Briquet's catalog, the star and anchor usually indicate the involvement of Venetian masters.[1] Looking at watermarks from the period, however, it becomes obvious that similar watermarks were used throughout northern Italy (and even in Salzburg) between 1530 and 1600. The only difference between them was a small curly line here and there. In order to achieve more precise dating, the matter of these curly lines must be clarified.

At this point in my investigation, I felt the need to seek expert advice, which I found in the person of Joe Nickell, who happened to be in Budapest as a guest of the European Skeptics Congress. Nickell has exposed a great many counterfeit antiques, academic bluffs, hoaxes, books, and texts pretending to be old, and he has published several books about these projects and his research methods.[2] The best of these, in my view, *Pen, Ink, and Evidence*, despite what the title suggests, is not merely written for enthusiasts of writing tools and types of ink. It also caters to so-called document detectives, people who want to learn about the often very simple methods that help expose counterfeit documents or verify authentic ones by examining the ink, the paper, and the writing on the document.

Among many other special documents, Nickell has examined the diary of Jack the Ripper, the late nineteenth-century London serial killer who mutilated his victims with a knife. The diary was unexpectedly found in 1991 and made headlines the world over, for it finally opened a way to identify the murderer, something that had not been possible for a century. Nickell and his colleagues, however, proved that the sensationalist document was a modern counterfeit by examining the text, analyzing the content, and examining the paper and the ink in a laboratory setting. As a result, the publishing company halted the printing of the incomplete diary at the very last moment. Nickell has received a lot of publicity by scientifically exposing counterfeits or verifying authentic historic sources; the hero of the Hollywood movie *The Reaping* (starring Hilary Swank) was based in part on Nickell.

When we met at the congress, I only had a couple of minutes to get Nickell interested in the codex, but he immediately agreed to accompany me to the Academy's library the following day to have a look at the manuscript. When we examined it together, he first identified the watermark and then analyzed the paper and the ink as well as he could outside the laboratory. He concluded that the paper is an early, handmade, so-called ribbed paper that reveals vertical and horizontal structures. This and the presence of the watermark indicate an old book.

So what? you may ask. What does the fact that the paper was indeed made in the sixteenth century prove? The codex could still have been written centuries later. This sounds like a fair argument, but we must remember that paper was a pricey commodity in the sixteenth century, and it was fairly rare that a complete book made of paper remained blank for hundreds of years. It is more likely that the codex was bound and written on not long after it was made, presumably not later than the late sixteenth or early seventeenth century.

This hypothesis is supported by Joe Nickell's analysis of the script: it was written with a broad pen, which was common in the sixteenth and seventeenth centuries. The iron gall ink was widely used in that period, and the stains on the pages got there sometime after the text was written, since the ink is not smeared. All of these features argue in favor of authentic old writing.

Nickell thinks that the fact that the document contains a relatively small number of deletions and corrections indicates that it was copied from an already complete document. Finally, he disclaims my earlier theory of several hands. He argues that the differences in the appearance of the script are the result of the author's writing at different speeds, changing ink, or sharpening the nib. He also thinks that the drawings were created by the same person, and he dates them to the same time that the writing was made, since they are an organic part of the text. I had previously believed that there could have been two or three different people behind the codex, and others have argued for as many as twenty authors. When looking at the actual manuscript, however, and not only its facsimiles, it does look more likely that merely "one hand" is responsible for the whole book.

In a nutshell, Joe Nickell found no obvious sign that the writing is considerably newer than the paper itself, or that it is a counterfeit. To his expert eye, the writing looks as old as the book. Of course, we cannot be absolutely certain until laboratory tests have been carried out. Blank books made of paper did sometimes hide in libraries, albeit infrequently, and the enormous Batthyány library was one of them.[3] Whatever the truth might be, the *terminus ante quem* (the latest possible date for the codex) is 1838, while the *terminus post quem* (the earliest possible date) is 1530.

When did the codex become part of the Batthyány collection? As noted above, it probably entered the library much earlier than its public discovery in 1838. Following this track, I tried to exclude the possibility that the bookdealer and forger Sámuel Literáti Nemes had anything to do with the codex. Fortunately for us, the Batthyánys had made lists of their books several times over the centuries, so it seems likely that we should find some trace of the codex in those lists. The nineteenth-century scholars János Jerney and Kálmán Némäti believed that the "Hungarian prayers," a book similar in size to the Rohonc Codex, from the Batthyánys' 1743 list of books, could in fact be the Rohonc Codex, although the codex does not look Hungarian, and it's unlikely that it would have been identified as such. Moreover, it is doubtful that during a painstaking catalog-making procedure, a unique book like the Rohonc Codex would have been listed without any reference to its biblical pictures and extremely odd writing system. Since the codex was just as perplexing in 1743 as in 1838, it would be strange if it had simply been labeled "Hungarian prayers." But I have not been able to find any reference to a strange-looking codex in either the earlier lists of books or the published correspondence of the family.[4] Prayer books there are, in abundance, in the family collections from different time periods, but, interestingly enough, these are more likely to be in Latin, German, Italian, or French than in Hungarian. To date, we do not know of any written in an unknown language.

THE ROHONC SCRIPT

It looks like we are still flying blind when it comes to most of the information surrounding the codex. The right-to-left, top-to-bottom direction of the writing is obvious even to superficial inspectors, but there is no title page that would suggest the contents, and the leather binding is a later addition. An unusually large number of characters in the alphabet can be identified, but even listing these properly is hard: new characters crop up everywhere. There are no clues as to the language behind the system of symbols, as there are no obvious letter, word, or sentence boundaries, making it impossible to identify which sign stands for which letter, syllable, or word.

Putting these uncertainties aside for now, it is possible to make a few positive statements about the writing system and the contents of the codex. By strict logical reasoning, one can arrive at three distinct possibilities:

A. The Rohonc Codex is a text without any meaning; therefore it is impossible to solve.
B. The text is meaningful but cannot be solved for some reason.
C. The text is meaningful, and, at least theoretically, it can be solved.

These statements result in several further possibilities:

1. The codex is a hoax, which possibility leads to case A.
2. It is written in a natural language—leading to case C.
3. It is a cipher,
4. a stenography,
5. or an artificial language—leading to either B or C, as will be explained later.
6. Finally, the script consists of the notes of twins who grew up in isolation and who invented a language that nobody else understood (called cryptophasia in the literature)—leading to B.[5]

Coming back to possibilities no. 3 and no. 4 above, if the Rohonc Codex is a cipher or a stenography (a system of shorthand writing, discussed at more length in chapter 7), then it can also be:

3–4a. a writing composed of letters, syllables, or consonants—case C;
3–4b. a mixed system using symbols for both letters and codes—case C;
3–4c. a system based exclusively on codes, which can only be broken with great luck once the list of codes is lost. Similar to an artificial language (no. 5), it is case either B or C.

Although I could not rule out possibility no. 1 with complete certainty, I had already concluded that it would be very unlikely for someone to write 450 pages of made-up text for no specific reason. Looking neither valuable nor early Hungarian, the book was not meant to deceive book collectors.

Possibility no. 2 is even less likely, and not only because of the failure of those codebreakers (Viorica Enăchiuc, Mahesh Kumar Singh, and Attila Nyíri) who expected to find a natural language behind the code. More important, it is not typical that sixteenth-century natural languages of more than 150 letters survived in only one source.

Possibility no. 6 is a difficult one to disprove definitively. Theoretically possible, it is still improbable that the biblical text of the Rohonc Codex would be the result of cryptophasia. Since this (very remote) possibility would rule out the prospect of successful deciphering in any case, I will not follow it up.

Let us focus, then, on examining cases no. 3, no. 4, and no. 5. If we accept the thesis that the codex is a cipher, a stenography, or an artificial language, we are faced with a queer situation. Before 1838, the year in which the codex was discovered, no enciphering method existed that could withstand the efforts of modern-day codebreakers. There is an unusual contrast between the outdated encrypting method and the fact that, despite all efforts, it is still unbroken.

A strange link connects the three different cases, regardless of the different intentions that each one presupposes. Whether secrecy was intentional or not, the codebreaker's goal is the same: to use cryptanalysis to decipher the writing system. Perhaps the author did not intend to create a cipher but only a quick and easy writing technique (no. 4), or maybe a transparent language (no. 5) that would be easy to recognize and could be used to illustrate the truths of Christianity. It is all the same to us now that no character table or dictionary survives. We can still approach the text as if it were an encrypted system.[6]

If we happen to come across an early modern cipher, stenography, or artificial language created with a method similar to that employed in the Rohonc Codex, the rest will be easy. Until then, however, we had better look for a *crib*, a widely used codebreaking method. A crib is a word, name, or concept that, based on our assumptions about the content, the ciphertext is likely to contain. A twentieth-century encrypted war report, for example, almost always has a date at the beginning or end of the text, and the weather is discussed somewhere in between. An encrypted report from a sixteenth- or seventeenth-century envoy will most certainly start with a salutation. Aware of this, the codebreaker looks for a string of characters that is structured like a salutation, date, or weather report. The famous "grand chiffre" (great cipher) of Louis XIV, for example, resisted codebreakers for centuries because it assigned numbers to the component parts of words, 587 different ones in all. The breakthrough came in the 1890s, when the French military cryptanalyst Étienne Bazeries assumed that the component parts of the word *les ennemis* (the enemies) were likely to occur in encrypted war reports. Bazeries tried substituting these parts (*les, en, ne, mi, s*) with the numeral combination 124-22-125-46-345, which appears often in the text. When he took these word parts and tried them with other parts of the text, he saw that they worked there too. Identifying all 587 word parts was a tremendous undertaking, but the encrypted seventeenth-century sources could finally be read.[7]

If such a crib were found in the Rohonc Codex, at least some parts of the code could be deciphered, or the encoding method could be recognized. An excellent starting point would be the title page, where the name of the author, a title, and perhaps a date would be found. This method worked for the grandson of Simeone Levi, the nineteenth-century Italian Egyptologist. Levi had used a complicated syllable system to encrypt his biography, but it was broken by his curious descendant, who found a crib on the title page.[8] But the Rohonc Codex has no title page. Fortunately, however, the captions to the codex's illustrations, all depicting biblical themes, contain frequently occurring strings of characters. It thus makes sense to start the investigation there, looking for symbols that obviously refer to certain recognizable people, scenes, and events.

CHAPTER 4

THE JESUS CODE

THE ILLUSTRATIONS

As the script in the Rohonc Codex says nothing about its content or cultural background, the codebreaker must rely on what the drawings disclose. Fortunately, there are more than ninety of them. Most of them are separated from the text by a decorative border, but a few occur within the lines of text—three of these are decorative borders, and four, which fall at the end of certain chapters, look like a very long fish, or a snake or eel. The list of illustrations and the figures identified in them can be found in the appendix to this book.

The primitive, sometimes naïve, illustrations might be dismissed as childish scribbles at first glance. Careful study of them, however, reveals a great deal about the codex. Most of the pictures (especially those in the first half of the manuscript) depict inarguably biblical topics, almost exclusively connected to the well-known and often depicted events of the story of Christ from the Gospels: the Annunciation, the Magi with the Star of Bethlehem, Jesus entering Jerusalem, Christ before Pilate, Christ bearing the cross, the Crucifixion, the Resurrection, the Ascension, and the events following Christ's death.

The images that follow those illustrating the Gospel story are harder to identify. They usually depict a teaching "saint," most probably Christ himself, with the cross, and the halo on his head, talking to one or more disciples—a possible reference to biblical parables. This "saint" seems to be praying to the sun in many of the pictures, and the unusually long rays of the sun seem to touch him. In other pictures he is standing above and looking down on a city, presumably Jerusalem. Some of the pictures may refer to the book of Revelation, while others seem to depict the structure of the world.[1]

As for the numbering of the folios (and, consequently, of the drawings), I followed the recto and verso folio numbering that is usually applied in manuscript studies. However, the numbering of the folios, following the direction of the script, goes backward. I rely on the nineteenth-century numbering that is also used in the online and printed versions of the Rohonc Codex. That is, the last sheet, where one begins reading, is folio 1. Every left-hand page is designated a number starting from there, and the page opposite it is the reverse of that page—in other words, the verso of the previous one. Because the codex begins at what we normally consider the end of the book and reads right to left, this means that right-hand leaves (normally labeled *r* for *recto*) are labeled *v* (for *verso*), and left-hand leaves are labeled *r* (for *recto*)—the exact opposite of normal folio numbering—as shown below.

3r	2v		2r	1v		1r	

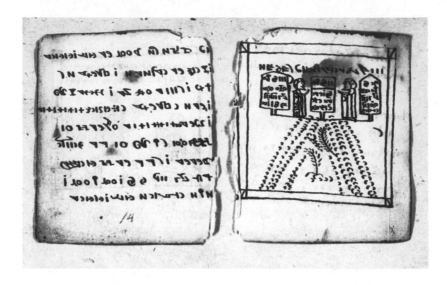

FIG. 4 | Unidentified scene, probably Moses with the Tablets of Law on Mount Sinai. The Rohonc Codex, Library of the Hungarian Academy of Sciences, MS K 114, fols. 14r–13v.

FIG. 5 | Christ entering Jerusalem and the cleansing of the Temple (to the right). The Rohonc Codex, Library of the Hungarian Academy of Sciences, MS K 114, fols. 15r–14v.

FIG. 6 | The Annunciation, in which an angel (presumably the archangel Gabriel) hands Mary an enormous lily, while Joseph and another angel converse atop a mountain at right (16r); the sacrifice of a pigeon in the Temple (15v). The Rohonc Codex, Library of the Hungarian Academy of Sciences, MS K 114, fols. 16r–15v.

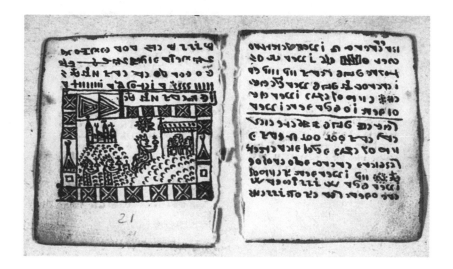

FIG. 7 | The Adoration of the Magi. The Rohonc Codex, Library of the Hungarian Academy of Sciences, MS K 114, fols. 21r–20v.

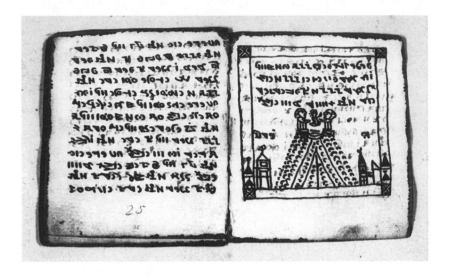

FIG. 8 | Possibly the Transfiguration of Jesus but more probably a baptism scene. The Rohonc Codex, Library of the Hungarian Academy of Sciences, MS K 114, fols. 25r–24v.

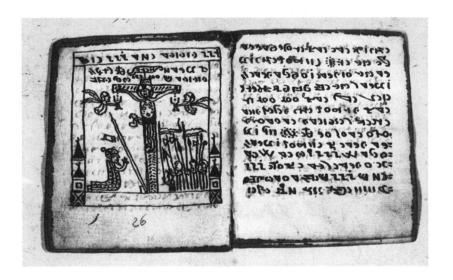

FIG. 9 | The Crucifixion of Christ. The Rohonc Codex, Library of the Hungarian Academy of Sciences, MS K 114, fols. 26r–25v.

FIG. 10 | Christ standing before Pilate. The Rohonc Codex, Library of the Hungarian Academy of Sciences, MS K 114, fols. 41r–40v.

FIG. 11 | Christ before Pilate wearing the crown of thorns. The Rohonc Codex, Library of the Hungarian Academy of Sciences, MS K 114, fols. 45r–44v.

FIG. 12 | Two altars, each with a priest, at the foot of the cross. The Rohonc Codex, Library of the Hungarian Academy of Sciences, MS K 114, fols. 52r–51v.

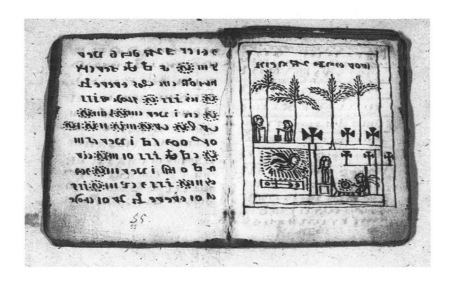

FIG. 13 | The events following the death of Christ: the *Noli me tangere* scene, where Mary Magdalene meets the resurrected Christ and mistakes him for the gardener (above); the empty tomb with the angel (bottom left); and John at the tomb (bottom right). The Rohonc Codex, Library of the Hungarian Academy of Sciences, MS K 114, fols. 55r–54v.

FIG. 14 | The Resurrection of Christ. The Rohonc Codex, Library of the Hungarian Academy of Sciences, MS K 114, fols. 59r–58v.

FIG. 15 | Apocalyptic scene (left) and the Tree of Knowledge, the four rivers, and Christ standing between two people (right). The Rohonc Codex, Library of the Hungarian Academy of Sciences, MS K 114, fols. 79r–78v.

FIG. 16 | The structure of the world with the spheres and heavenly bodies. The Rohonc Codex, Library of the Hungarian Academy of Sciences, MS K 114, fols. 83r–82v.

FIG. 17 | A crowned person praying to the sun. The symbols in the sun are reminiscent of the Hebrew tetragrammaton. The Rohonc Codex, Library of the Hungarian Academy of Sciences, MS K 114, fols. 134r–133v.

Although the drawings are sketchy, their details are likely to lead to certain conclusions. At this point in my research, I asked again for expert help; this time, I consulted art historians. They called my attention to the fact that the figures depicted are typical of the time when the codex was allegedly written, even though their naïve style cannot be used directly as a clue (see figs. 4–17). Details like the form of the halberds (shafted weapons with axe-like blades) and the architectural features of the temples (steeple, nave, apse) suggest the sixteenth and seventeenth centuries, and most probably the central European region. The architectural elements rule out Near Eastern, Turkish, or Arab origins. Some of the drawings include indications of Orthodox Christian culture (for example, a third horizontal crossbeam at the base of the cross), while others suggest Islamic culture (a crescent at the top of the temple). Together with the overwhelming presence of Western Christian symbols, these features support the theory of a central European (perhaps Transylvanian, Romanian, Balkanic) origin.[2]

Although one can exclude the Near East as a place of origin but not western Europe, this reduces the number of possible languages that were the codex creator's native tongue to a certain degree. Some of the candidates are Latin, German, Hungarian, southern Slavic, and even Romanian. We must also consider Hebrew and Ottoman Turkish, and not only because of their strong presence in central Europe but also because both languages are written right to left.

The illustrations can be helpful in this quest because of the inscriptions they contain. Identifying their content can be the first step in discovering what letters or syllables the particular symbols stand for. These cribs, as we have seen, are dear to the cryptologist, as they may provide an entry point into the code system. If it is obvious what such an inscription *should* mean, it is easier to figure out *in what ways* it represents that given meaning. A correctly identified inscription may play the same role that the Rosetta Stone did in deciphering the Egyptian hieroglyphs: the proposed meaning of the complete inscription was the basis for identifying the meaning of each particular symbol.

I did my best to identify such cribs so that I could use them to draw certain conclusions. I copied all of the illustrations that contained inscriptions into one single file and stared at them for hours. I arranged the repetitive symbols next to one another. The four-character sign at the top of the cross (see fol. 26r in fig. 9) is such an obvious starting point as a crib that even a beginner codebreaker with superficial iconographic knowledge would try to match it to the standard INRI seen in similar pictures of the crucified Christ.

A logical idea but, sadly, there are two immediate problems with it. One is that the first and last characters of the sign that should correspond to INRI are not identical. The two symbols that should represent the letter *I* in the text "Iesus Nazarenus, Rex Iudaeorum" are different. The other is that the whole inscription differs slightly from the INRI sign on another cross elsewhere in the codex (see fol. 52r in fig. 12). These two factors suggest that the same letter may be represented by different symbols in different places, something that is not at all surprising given that the number of symbols greatly exceeds the number of letters in the alphabet.

There is another promising pair of characters seen in many illustrations above the head of Christ that most certainly refers to him. It can be seen in the picture where Christ is standing in front of Pilate, above whom is another double symbol (see fol. 40v in fig. 10). We have every reason to suppose that these two double signs stand for the names of these two actors. The signs that stand for the name of Christ can be found scattered all over the book. The symbol combination indicating Pilate also occurs frequently on the pages preceding and following the illustration on fol. 40v, pages that are very likely to be telling the part of the Gospel story that mentions Pilate, but it occurs nowhere else in the codex. This makes sense, since Pilate is present neither in Christ's childhood nor in the events following his Resurrection. Identifying the name of Christ solves another INRI problem: one part of the double sign is the same as the first character in the INRI sign. This could represent the name of Christ in the abbreviation on the cross and would explain why the first and the last symbols are not identical: these are particular symbols that stand for the name of Christ in the codex.

The fourth crib is presented in recurrent illustrations where Christ is depicted surveying Jerusalem. Here, there is always a symbol written above the city. This symbol, by the way, looks very similar to the last character of the INRI sign. There is good reason to suppose that this symbol stands for the letter J, as in Jerusalem or Judaeorum (fol. 111v).[3]

More obscure is the meaning of the double symbol that appears at the top of every page between folios 31r and 50r, like a chapter title, section heading, reference, or some kind of page numbering. This is peculiar because its first part, which looks like the letter N, is the same as the first character in

Christ's name, whereas its second one, which resembles a T with an arch, often appears as the third character following the two that make up the name of Christ.

The illustration of the Annunciation (see fol. 16r in fig. 6) is another promising place for cribs. The barely legible string of characters should somehow correspond to the text "Ave Maria, gratia plena" (Hail Mary, full of grace) or at least "gratia plena," according to all other depictions of the angelic visitation.

Most codebreakers are pleased if they can identify even the beginning or end of a text with any certainty. Knowing that a character combination occurs at the beginning or end of a sentence doesn't seem like much, but many codebreakers have built on such humble beginnings to solve a complete code system. In the case of the Rohonc Codex, things were looking promising at this stage of my work. Having examined all the codex illustrations that contained inscriptions, I had found a substantial number of character strings that clearly begin or end a section of the text. Identifying the topic of a given picture could help us find the meaning of these strings.

I will avoid elaborate technical explanations here. Suffice it to say that we can be certain that the symbols are used consistently in the codex, and this certainty leads to several more or less valid conclusions. At the very least, it proves that the symbols in the text of the codex were not recorded in an unintentional, random, or meaningless way.

At the end of chapter 2, I mentioned two researchers who spent years trying to crack the Rohonc Codex. Gábor Tokai and Levente Zoltán Király started out from this very point, applying similar principles. They identified the characters discussed above, arriving at sometimes similar, sometimes different conclusions. However, they were more far-reaching than I, and thus far they have been the most successful in interpreting the strings of characters in the codex using rational methods. Tokai was able to transcribe complete sentences and to identify the numerals and the signs for the four evangelists. I discuss many of his results in more detail below, but he concluded, in short, that the person who created the Rohonc Codex tried to grasp language at the conceptual level. In other words, most of the signs in the codex do not merely correspond to letters but refer to complete words or concepts.[4] Király agrees, though initially he was more cautious about assigning actual meanings to the strings of signs. He used methods similar to Tokai's in some respects and different in others, and his achievements cover such areas as the syntax of the code and the structure of the text. In addition, like Tokai, Király managed to number correctly the loose pages of the book and to identify the numerals.[5] These results, along with an article the two men wrote together in 2018, in which they offer a solution to the codex, are discussed in chapter 9.

But first we must return to the problem of cribs in the illustrations. What type of document could contain exactly these pictures and the words identified so far in this exact arrangement and in this structure?

APOCRYPHAL GOSPELS

If we reject the hypothesis that the visual material is there only to deceive would-be codebreakers, then the illustrations and the symbols referring to the names of Jesus and Pilate prove clearly enough that the text is either a Gospel or at least a retelling of the events described in the Gospels. It soon becomes obvious that the text of the codex is different from any of the four canonized versions of the Gospel story, despite the typical iconographic features of the illustrations. The scene of Jesus entering Jerusalem, followed directly by his chasing the money changers out of the Temple, refers to the Gospel of Matthew; the offering of the pigeons refers to Luke; and a number of other scenes, such as Mary Magdalene's meeting the gardener, refer to the Gospel of John and Revelation. The names of Pilate and Jesus do not occur as frequently in the Bible as they do in the script of the Rohonc Codex. Neither the synoptic Gospels nor the Gospel of John mentions these two names so frequently. The Rohonc text must be a more detailed description of the life and death of Jesus than the Bible itself contains.

It is tempting to believe that we are dealing with an Apocryphal Gospel here. There are almost forty different versions of the story and teachings of Christ and other Gospel actors, of which only four became part of the biblical canon. Those that were left out, among them the Gnostic Gospels, were typically written a little later than those four and were deemed heretical by the decision makers of the early church, although these writings do not strike the superficial modern-day reader as especially wild or threatening. The Gospels that were left out of the Bible first drew the attention of scholars at the end of the nineteenth century, and then in the mid-twentieth, with the discovery of the famous Nag Hammadi scrolls (not to be confused with the equally significant Dead Sea Scrolls, which contain early copies of Old Testament books, not apocryphal New Testament documents).[6] Although these alternative biblical stories were banned by the church, as they were on the famous *Index librorum prohibitorum*, the list of prohibited books, they spread on several cultural levels. There are numerous traces of them in European folklore—for example, the familiar Christmas scene in the cave of Bethlehem, which is not actually from the official Bible, either.

There were about thirty New Testament Apocrypha in medieval Russia, for example; a fourteenth-century poem in Czech was based on pseudo-Matthew and Thomas Gospels, and Apocryphal topics were abundant in the sermons of pastors, too.[7] Apocrypha survived not only in Arabic, Armenian, and Ethiopian but even in Old Bulgarian (from the ninth and tenth centuries) and in secular Slavic variations.[8] Particularly interesting is the third-century *Pistis Sophia*, a book written in Coptic that was discovered in the eighteenth century, a simpler version of which was also unearthed at Nag Hammadi. It tells the story of the eleven years that Christ spent teaching on earth after his Resurrection. This aligns neatly with the illustrations of the Rohonc Codex, as they depict stories from the Gospels, followed by the teachings and parables of Christ.[9] The Rohonc Codex could be an encoded version of any of these sources, or, to hazard an even more daring proposition, it could even be an Apocryphal text that was not among either the texts found in Egypt in 1886 or the Gnostic Gospels unearthed in 1945 at Nag Hammadi.

Sadly, however, we must scratch from the list of possibilities the Apocryphal Gospels that have been found so far, at least the versions known today, for they, even the so-called Gospel of Pilate, describe the meeting between Jesus and Pilate much more concisely than the Rohonc Codex seems to do.

What this means is that we are left with the thrilling possibility of dealing with an undiscovered Gospel. Unfortunately, this is not a likely scenario. To start with, chances are very slim that an uncanonical Gospel that was written in the first, second, or third century and that remained completely unknown survived in no other copies than a seventeenth-century secret codex from central Europe. Even if this were the case—if the codex is really the sole surviving copy of an undiscovered Gospel—we have no text with which to compare it, and thus no clues as to how to crack its code.

Let us consider the possibility, then, that this seventeenth-century Gospel variation is not an ancient text written in the first two or three centuries after Christ but a later book that was rejected for some reason and thus became Apocryphal. There are precedents—for example, there are texts considered sacred by the medieval Chatar and Bogomil heretics that the official church did everything in its power to suppress and extinguish. These texts introduced their own teachings through the usual biblical stories and the words of Christ to his disciples.[10]

I tried but failed to match the surviving Chatar and Bogomil Apocryphal Gospels to the text of the Rohonc Codex, comparing both the signs for Christ and Pilate and the repeated text strings that begin and end chapters. Major parts of these documents are available not only in manuscript but also in printed form in the libraries, so I was able to read piles of them— and still, no luck. But I was not discouraged. I still held out hope that there might have been one small sect in the age of the Reformation that was not accepted by the Catholic, Lutheran, Calvinist, or Unitarian churches and that put together a secret prayer book, which it tried to keep hidden from the mainstream religious authorities. While the illustrations do not suggest any unorthodox or heretical dogmas, we cannot with certainty say the same of the text—after all, most of the theological disputes that caused bloody wars against sects and small churches in the sixteenth and seventeenth centuries seem insignificant today. Many reformed Christian sects believed in stripping away the accretions of centuries of imposed dogma and returning to the authority of scripture alone, so it is no surprise that a prayer book from this era should contain primarily biblical material. I could not rule out the possibility, then, that the Rohonc Codex is an Apocryphal New Testament text that contains the teachings of a little-known sixteenth-century sect, formed around the same time that the codex was written. If this were the case, it would be no wonder that we do not find any other copies of this prayer book. Such a hypothesis is not preposterous, far-fetched, or outrageous, and in the course of my research I kept it alive as one possible answer to the puzzle.

From time to time, however, I temporarily set aside this thrilling possibility to forge more prosaic, perhaps more realistic theories. The Rohonc Codex might, for example, be a simple *Vita Christi* (life of Christ), a very popular genre in late medieval and early modern times designed to satisfy the curiosity of those who wanted to know more about the life of Jesus than they could find in the Bible. In terms of content, *Vitae Christi* were carefully tailored to official church teaching, yet they offered a more detailed description of the sayings and events of Christ's life than the Gospels contain.

Naturally, this kind of text could describe the dialogue between Christ and Pilate as lengthily as Mikhail Bulgakov does in *The Master and Margarita*.[11] The most popular of these writings was *Meditationes vitae Christi* (Meditations on the life of Christ).[12] For a long time it was believed to be the work of the famous medieval theologian Saint Bonaventure (1221–1274), but we now know that it was written by the less famous Ludolf the Carthusian (ca. 1300–1377), who exploited this supposed authorship.[13] The extremely widespread *Biblia pauperum* (Paupers' Bible)[14] satisfied a similar need, as did the so-called *Songs of Christ*, the epic songs sung by the common people about Christ's birth, life, death, and miracles, which were half biblical, half Apocryphal in origin. There is in fact a bewildering array of texts that could have served as the basis for the Rohonc Codex.

The situation would be similarly prosaic if we were dealing with a simple Book of Hours, which is an even more likely scenario, since this genre was even more common in the sixteenth and seventeenth centuries—almost as common as the Bible itself.[15] These books included prayers, psalms, litanies, Gospel stories, and parables in various arrangements and were similar to breviaries, the prayer books of the clergy. The only difference was that Books of Hours served laypeople, and their arrangement of texts was much less tightly controlled by church authorities than in the clerical version. Among all the genres that I know of, the structure of the Book of Hours—which usually starts with sections from the Gospel story, followed by prayers—most closely resembles that of the Rohonc Codex. Since they mostly serve personal purposes, they are usually small pocket books, like our encrypted codex. The simple engravings in the printed versions of sixteenth-century Books of Hours are similar to the Rohonc illustrations as well. They have the same odd-looking frames, and the drawings are also relatively simplistic. The crescent may also appear here and there on church drawings in a Book of Hours,[16] as do the sun and moon, a phenomenon that earlier scholars believed to be unique to the Rohonc Codex. So why would we discard the idea that the illustrations in the codex imitate the engravings of an early modern printed publication? But an obvious question arises, and a baffling one: why would anyone encrypt such an innocent type of book?

While I have not been able to answer this question, the assumption that the codex is a prayer book, or a simple text similar to a prayer book, offers us hope, because it explains the very large number of repetitions, which have driven many a codebreaker to the brink of madness. As we saw earlier, cryptanalysts are usually thrilled to find any kind of pattern or repetition in seemingly meaningless strings of characters. These patterns and repetitions were the key to even World War II–level cryptanalytic methods.

The problem is that the Rohonc Codex contains too many of these repetitions. It makes you wonder what kind of text can repeat this many words, concepts, or sentence parts within the space of a single page. What is even more striking is that several paragraphs and sections begin, and even end, with almost the same text. One example is the string of characters that appears on folio 14v, at the bottom of page 64v, and at the top of folio 65r, and bear in mind that the text goes from right to left, so that the first word falls at the right end of the line and the last word at the left end. If the codex were a hoax, how would we explain these repetitions? Did the author's creativity flag from time to time, so that he chose to simply copy passages from elsewhere in the book in a rather uncreative fashion?

But if the codex is a prayer book full of litanies, the endless repetitions are no longer surprising. Prayer books are known for their many repeated textual units. Those who read a prayer that they understand do not even notice these repetitions (e.g., *Sanctus, Sanctus, Sanctus, Dominus Deus Sabaoth*), but the same repetitive structure in an unknown text looks unnatural, as we do not know the meaning of the words. Seeing the same words immediately after the pictures (that is, at the beginning of a paragraph) in the Rohonc Codex should come as no surprise either. Books of Hours, for example, often have the word *Domine* or *Deus* as the first word that follows a picture. Gábor Tokai and Levente Király assume that the text in the codex explicitly refers to certain parts of the Bible, and thus the strings of characters following the pictures could well be biblical references.[17]

And we still have not run out of possible scenarios. Once the visible features of the text are taken as a starting point, the direction of the writing should obviously be considered as well. Although many encrypting methods use right-to-left writing, this is not a very effective encryption strategy, for it fools no one, adding perhaps five minutes to the work of the cryptologist. Many historical figures, among them Leonardo da Vinci, are known to have used mirror writing, but this method only succeeds in confusing uneducated servants or post office apprentices. Having spent hours with a handy computer software program that mirrored the pages of the Rohonc Codex (in all directions and in every possible combination), I realized that this was a dead end. So I looked for other explanations.

There are two possible explanations for the unusual right-to-left writing in central Europe: the influence of Hebrew or the influence of Arabic/Turkish languages. In the case of Hebrew, there is a fairly common text titled *Toldoth Jeschu* (The life of Jesus), which depicts Jesus not as the Son of God but as a great magician. But this view of Christ and the Christian faith is not at all reflected in the illustrations of the Rohonc Codex. Furthermore, the Jewish Jesus stories, however offensive to Christians, were quite openly shared, and there is no reason to think that anyone would need to encrypt such a text.

There were, however, other religious denominations that mixed the Jewish and Christian cultures. They considered the former a fruitful and complementary forebear of the latter, as mainstream Christian denominations still do today. Several antitrinitarian Christian sects were labeled "Judaizing" in Hungary during the sixteenth century, but these movements were actually quite different from one another.[18] They shared one thing in common, though: they all knew Hebrew culture and language well, and they took it seriously. Many of these sects could potentially have penned a Christian religious text in Hebrew. What is less likely is that they would translate a Gospel and a long supplementary text with prayers into Hebrew, since these "Judaizing" groups (whose members kept the Sabbath) were more interested in Old Testament and rabbinic literature. However, it was during my research on the Sabbath keepers that I stumbled upon the following story, which is vaguely reminiscent of the Rohonc Codex. Martin Seidel, a Sabbath-keeping author, wrote a secret book titled *Origo et Fundamenta de Religione Christianae*. This book was secretly carried from one place to another, after the covers had been removed and the pages glued back into it in reverse order for security reasons. Despite these efforts, the soldiers of the Prince of Transylvania found and confiscated the manuscript.[19]

The other reason why someone might write from right to left at that time was the strong presence of the Turkish language in central Europe. Most of southeastern Europe and one-third of Hungary was under Turkish rule in the sixteenth and seventeenth centuries. The influence of Turkish culture, and the Arabic writing accompanying it, is not to be underestimated. Turks, unlike Judeo-Christians, had every reason to encrypt Christian writings. While the Ottoman Empire had a relatively and pragmatically tolerant approach to the Christians whom they had oppressed, not even bothering about their prayer books, it was very intolerant of its own people who renounced the Islamic faith and embraced Christianity, particularly those who would go so far as to write a Christian prayer book or join a Christian sect.

This theory also has its weak points, however. For whom would such an apostate of Islam write a secret Christian prayer book? It is easy to imagine that an individual might become a secret Christian convert, much harder to picture a whole group of friends and followers who would become the readers of such a secret book. Were they all Turkish people who converted to Christianity at once, while outwardly remaining servants of the Ottoman Empire? Possible, but not very likely.[20]

The reverse was also possible: there were people who were born into Christianity but converted to Islam, often for political reasons, in an attempt to escape the persecution of their homeland and seek freedom of thought. Others claimed to have been persecuted but were in fact petty criminals on the run. And then there were those who were captured, pretended to return to the Islamic faith, but continued to practice Christianity in secret.[21]

I spent many hours trying to identify a specific Gnostic, Chatar, or Bogomil Apocryphal Gospel, or perhaps a prayer book, Book of Hours, breviary, Bible commentary, Christian or Jewish Jesus story, or maybe the writing of an ex-Muslim Christian, or an ex-Christian Muslim, or a Christian who first became Muslim but then reverted to Christianity. Or anything else that could have been the text of the Rohonc Codex. Although I had no idea whether the recurrent strings of characters stood for words, letters, or parts of sentences, I tried to project them onto extant texts to see if they matched. They did not.

Similarly, I browsed through large printed and online collections of western European, Greek Catholic, Old Church Slavonic, and Armenian paintings, drawings, frescoes, and book illustrations to find analogues to the drawings in the Rohonc Codex.[22] I was hoping to discover the meaning of the text within the pictures, or to identify a few peculiar iconographic details. I had been able to do the latter (more about that in the appendix), but I must admit that I failed to achieve a breakthrough.

In fact, this research only made things look even more complex. As possible research tracks split into several subtracks, the number of options constantly grew. Having studied the colorful central European region, I ended up with a Babel-like list of the possible languages that might have been the codex writer's native tongue. Hebrew, Turkish, Bosnian, Old Church

Slavonic, Serbian, Romanian, Hungarian, and even German all had to be considered, and two of these, Hebrew and Turkish, use basically consonant alphabets. I realized that I needed to find another way to limit the number of possible languages.

I decided to focus on the characters in the Rohonc Codex and determine whether they stood for letters, consonants, syllables, or concepts. In other words, I needed to discover the principle on which the coding system was based: was it a special enciphering method, a stenographic strategy, or an artificial language?

CHAPTER 5

WRITING IN CIPHERS AND CODES

THEORY AND PRACTICE

We would not have to waste a lot of words on the history of cryptography before 1400 if it were not for the Arabs.[1] Most of the cryptographic methods in Western Latin and later in national languages remained on the level of simple substitution until late medieval times: every single letter of the plaintext was replaced by a corresponding numeral, letter, or symbol. As the ciphertext was being constructed, the scribe took the letters from the plaintext one by one and wrote their corresponding symbol in the encrypted text.[2] This method assigned a single string of symbols, numerals, or letters to the alphabet; in other words, it used a single code alphabet to cipher the plaintext. This process is thus called a monoalphabetic cipher (see fig. 18).

Monoalphabetic ciphers are rather vulnerable. It may seem at first that in the case of a twenty-two-letter alphabet, the codebreaker must choose from 22! (twenty-two factorial)—that is, $22 \times 21 \times 20 \ldots \times 3 \times 2 \times 1 = 1{,}124{,}000{,}000{,}000{,}000{,}000{,}000$—possibilities, which is a highly time-consuming task, almost impossible without the help of a computer. Fortunately, the work of the codebreaker is not this difficult. Although there is a range of mathematical methods available with which one can radically

FIG. 18 | The monoalphabetic cipher of Ludovicus de Berlaymont, archbishop of Cambrai (1570–96). From Aloys Meister, *Die Geheimschrift im dienste der päpstlichen Kurie von ihren Anfängen bis zum Ende des 16. Jahrhunderts* (Paderborn: Ferdinand Schöningh, 1906), 271.

narrow down the number of possibilities to break this type of cipher, in reality many people use simpler ways to decipher monoalphabetic substitutions. They look for a prominent pattern in the flow of characters and assign it to the most frequent syllables of the supposed language of the plaintext. An excellent example of this is the "cipher challenge" that Simon Singh published in the original edition of *The Code Book*, his book on the history of ciphers. Singh offered £10,000 to the first person or team who could decipher ten coded messages. Having worked together for one year, a group of amateurs and professionals from all over the world finally solved the challenge in October 2000. Reading their published codebreaking methodology, we know firsthand that a clever codebreaker, instead of using difficult computer-assisted algorithmic methods, can apply pen and paper to solve the first few (simplest) steps of the challenge.[3]

This quick method, however, sometimes fails to yield results. This might be because the ciphertext has no word boundaries, because the language of the plaintext is not known, because spelling mistakes were deliberately inserted into the text, because only identical vowels were used, or simply because the text offers no easily identifiable patterns. In such cases, a second method, frequency analysis, is attempted. This is a down-to-earth method in which the codebreaker counts each character of the ciphertext and tries to match the most frequent ones to the most frequent letters of the supposed language of the original text. The reliability of this method is based on the fact that languages are strongly characterized by the frequency of their letters, a feature that tends to be constant in every text written in the given language. All of this is only true, of course, if the cipherer does not intentionally make spelling mistakes or use technical terminology (military jargon, for example, uses a smaller number of definite and indefinite articles). In modern English, the letters E T A O I N S H R D L U are the most frequent, in that order, with z being the least common. In French, the most common are E N A S R I U T O L D C; in German, E N R I S T U D A H G L; in Italian, E I A O R L N T S C D P; and in Spanish, E A O S R I N L D C T U, while the Hungarian list starts with the letters E A T L N. The relative frequency table of any language can be easily created by counting all the letters on half a page of text or, even more simply, by downloading a ready-made graph from the internet.

Because, in a monoalphabetic cipher, identical letters are substituted for identical ones, and different letters of the plaintext stand for different characters in the ciphertext, it takes only a few attempts to match the tallest columns in the frequency chart of the original language to the corresponding columns in the frequency chart of the ciphertext. The longer the encrypted text, the more precisely it can be fitted to the frequency chart of the original language—a half-page sample is usually sufficient for this operation. There is an abundance of internet sites and software programs that will automatically do the frequency analysis of the text we upload.[4] This assigning process almost never works automatically, of course, and it takes several attempts to successfully match the string of the most frequent characters in the code to the string of the most frequent letters in the original language.

In addition to studying the frequency of letters, advanced codebreakers also analyze the frequency of bigrams (sequences of two letters), trigrams (three-letter sequences) and digraphs (a pair of letters used to write one speech sound). The most frequent bigram in the English language is TH, and the most frequent trigram, not surprisingly, is THE. The most frequent digraphs are SS, EE and TT. In light of this kind of linguistic statistical data, monoalphabetic ciphers are not difficult to crack and cannot be considered secure.

As for ancient and medieval history, the contemporaries of Julius Caesar, Charlemagne, the medieval popes, and the northern Italian city-states apparently did practice monoalphabetic encrypting to some extent, but they did not employ frequency analysis. Even so, they rarely entrusted their diplomatic secrets to ciphers. Instead, noncryptographic methods were preferred for hiding important messages. Letters, for example, were often simply hidden in the messenger's clothes.

In the Western world, the birth of the science of cryptography in the strictest sense did not take place until the fifteenth century, when the simple monoalphabetic method was finally replaced by new strategies. This happened despite the fact that medieval Arab authors had already achieved significant results. As is well known, beginning in the twelfth century, Western culture already owed a great deal to Arab science in several fields. Europeans were familiar with the scientific achievements of the Arab world, the fundamental texts of which were being assiduously translated. The Arabs were equally fruitful in the fields of astronomy, mathematics, optics, logic, philosophy, alchemy, astrology, and practical magic, and their texts were read and translated into Latin by dozens of translators.

This enthusiastic reception of Arab culture did not characterize the science of cryptology, although medieval Arab writers knew how to decipher monoalphabetic ciphers using methods based on language statistics, and they were already designing newer and better encrypting methods. We have only recently recognized the real achievements of the Arabs since historians have started publishing Arabic Origins of Cryptology, a series containing the most important documents—that is, sources that were found in the manuscript collections of Istanbul.[5] In light of these texts, we can argue that cryptology, in its origins, is a truly Arab science. In other areas of science, such as mathematics, philosophy, and logic, the Arabs acted as *transmitters*, albeit very important ones. The science of cryptology was not inherited from the Greeks or Romans, to be developed further—it was created by the Arabs from almost nothing. There had been simple encrypting methods in the ancient Greek and Roman worlds, but Arab writers were the first to attempt to systematize the methods of *decrypting* ciphers. This is how the actual science of cryptology was born, covering both cryptography (writing in cipher) and cryptanalysis (codebreaking, deciphering without a key).

The first of these Arab experts was a scholar from Baghdad, also prominent in the fields of philosophy, mathematics, and optics, the famous al-Kindi (ca. 801–873). Twelve hundred years ago, al-Kindi was already familiar with the method of letter-frequency analysis for breaking ciphers. In the ninth century, he could easily have broken the kind of monoalphabetic codes that the Western world used in diplomacy as late as the fourteenth century (and occasionally as late as the seventeenth). During the next half millennium, Arab cryptology flourished in the life and works of scholars, poets, and linguists from Damascus and Cairo, as seen from the handbooks of Ibn Adlan (1187–1268), Ibn ad-Durayhim (1312–1359), al-Qalqashandi (1355–1418), and others. These authors describe a variety of code types in detail, most of them basically monoalphabetic. For example, they suggest assigning the letters of a different language (Hebrew, Greek, Mongolian, Armenian, Persian, etc.) to the letters of the Arabic alphabet, so that the text is written in Arabic but with foreign characters. Alternatively, they could make up their own imaginary system of characters. Al-Qalqashandi mentions a procedure in which two Arabic letters correspond to a letter of the plaintext in such a way that the numerical values of the two letters equal the numerical value of the plaintext character (each letter of the Arabic alphabet has a numerical value). These methods are more progressive than the simple monoalphabetic substitution and do not yield their secrets to a simple frequency analysis.

It is far more important for our purposes, however, that Arabic handbooks contain cryptanalysis, the art of codebreaking. Cryptanalysis involves the careful study of the plaintext's language: the frequency of letters, letter pairs, letter combinations, and an analysis of typical word patterns. The authors suggest examining which letters occur together regularly and which ones are never used in combination; which letters typically begin words and which never do; which ones are usually doubled in an Arabic text; and which characters could signal word boundaries in a ciphertext that otherwise does not indicate word boundaries. Frequency analysis, nonetheless, remained the main tool of cryptanalysts.

Although these achievements allowed the Arabs to advance much further than Europeans in this field by the fourteenth century, their dominance declined in the following centuries. Western cryptography took off, thanks in part to a number of theoretical studies and in part to the practicing cryptologists of various Italian diplomatic services. Powerful Italian political centers soon employed their own codebreakers. The best known of these were the Argentis; Giovanni Battista Argenti and his nephew Matteo were employed by the pope. The Argentis were not only practicing codebreakers; they also wrote handbooks on cryptology in which they offered detailed advice on how to make an encrypted plaintext more difficult to decode and how to make a series of signs resist frequency analysis.

A good indicator of the growing prestige surrounding cryptology is the fact that these codebreakers were highly regarded, that they were separated from scribes working on simple, mechanical ciphering, and that assistants were often assigned to help them. In both the Papal Court and the Republic of Venice, this field was institutionalized around 1540 with the establishment of an official deciphering office. In Venice, the codebreakers' room was in the Doge's Palace, above the secret chamber. The codebreakers were not to be disturbed in their work, and legend has it that they could not leave their room until they had broken the incoming scripts.

Many leading scholars of the fifteenth, sixteenth, and seventeenth centuries published manuals on the developing art of cryptography.[6] When we consult the handbooks of Leon Battista Alberti,[7] Johannes Trithemius,[8] Giambattista della Porta,[9] Gustavus Selenus,[10] Blaise de Vigenère,[11] and the Englishman John Falconer,[12] we are at first astounded by the great array of methods available to cipherers in early modern times. Some of these were elementary operations based on monoalphabetic substitution, or on transposition (mixing up the letters of the plaintext), ciphers that could easily be broken by frequency analysis or by a vowel-identifying algorithm.

But these writers described more complex, so-called polyalphabetic ciphers as well. In these ciphers, letters of the plaintext are replaced by characters selected from several different code alphabets. As we encode one letter after the other, we switch code alphabets according to a simple system. The famous fifteenth-century architect Leon Battista Alberti (1404–1472) initially employed this method of using one alphabet to code the first few letters of the plaintext and then switching alphabets every few words. This polyalphabetic method was further developed by Trithemius, who prepared a whole table, in which series of complete alphabets are copied along horizontal lines, but each alphabet is shifted by one letter in each row (the first starting with a, the second with b, and so on). During the encoding process, he changed alphabets after *each letter* of the plaintext. Naturally, there are even more sophisticated methods. We can take a code word, for example (in honor of the inventor let it be *alberti*), and code the consecutive letters of the plaintext according to the alphabets starting with the letters of the word *alberti*. We look up the code sign for the first letter in the row that starts with the letter *a*, the second letter in the row starting with the letter *l*, and so on, till the seventh letter in the row starting with the letter *i*, and then we start again: the eighth letter is looked up in the row that starts with the letter *a*. We follow this method through the letters of the word *alberti* again and again, until we are finished with the coding.

This method was so good that it was only very occasionally broken before the nineteenth century. The main strength of this method, compared to monoalphabetic systems, is that it markedly raises the level of entropy in the text.

What is entropy in this context, by the way, and why does it need to be increased? Simply put, entropy is a measure of disorder, and mathematicians define the entropy of X using this equation:

$$H(X) = \sum_{i=1}^{n} p_i \log_2\left(\frac{1}{p_i}\right)$$

If this seems hard to grasp, consider the following. Every text written in a natural language contains strong patterns. In Latin (and in many Latinate languages), for example, the letter q is always followed by the letter u; *que* is very common, but *euq* and *uqe* do not exist, and *equ* is rare. In English, *h* often comes after *t*, but *t* rarely follows *h*. *The* occurs very frequently, but *eth* does not. In every language, the number of existing letter combinations is relatively restricted compared to the number of all such possible combinations. The stronger the structure, the greater the order in a given language, and the smaller the entropy (or disorder). Greater order, nevertheless, means greater predictability of the words in a given language. Even simple smartphone software can predict the second part of the word you are texting. The more predictable a system, the less secret it is and the easier it is to decipher, because once part of the text is decoded, the rest becomes easier to guess.

Even though monoalphabetic ciphers replace every single letter of the plaintext, the ciphertext will display the same patterns as the original text. Should the combination 22-17-46 occur relatively frequently with a space before and after it, there is a very good chance that this sequence stands for the letters of the article *the*. A polyalphabetic cipher mixes up the characteristic patterns of the base language, concealing its typical structures. Each instance of the word *the* in an English plaintext will look different in the coded version, for these three letters will be coded according to three different alphabets each time the word occurs. Admittedly, in a long enough sample text, if the code alphabets are used in strict rotation, we are likely to find two instances of *the* enciphered in the same way—an idea that served as the basis for the ingenious method of Charles Babbage (1791–1871), who was the first to finally break a polyalphabetic cipher. In general, however, polyalphabetic ciphers are much more difficult to break because they raise the level of entropy, or disorder.

Let us demonstrate this with an example.

There are seven instances of the letter *i* in the plaintext "In politics stupidity is not a handicap," each of which would be substituted by the same letter in a monoalphabetic cipher. In a polyalphabetic cipher, this strong pattern would disappear; the level of disorder, or entropy, in the text would increase. The ciphertext might look like this: *jp ssqoaqlc dfhdxtzlr cn jkq y gbpgmhgk.*

The text of the Rohonc Codex displays a particularly high level of structure and contains patterns that recur frequently (even line by line). In other words, its level of entropy is very low, even lower than that of a natural language, so we can rule out the possibility of classic polyalphabetic encryption. One less possibility to consider!

What else can be found in the major sixteenth-century cryptography handbooks that might help us crack the Rohonc Codex? One popular approach in these monographs is syllable methods, which assign a character to a syllable of the plaintext, double characters to letter pairs, perhaps a combination of three characters to a sequence of three letters in the plaintext, and so on.[13] In addition to cipher methods in the strictest sense, a number of other techniques for hiding messages are also discussed, usually methods used in the realm not of cryptography but of steganography, the art of concealing messages, often by embedding them within another message. For example, only certain letters in a text would be read, and the rest would be left out. Gustavus Selenus discusses such methods at great length, offering examples in which only the first letter of each word should be read, or only the first two, or every first letter but excluding the first and last words of each sentence.[14]

Having been introduced theoretically to such a rich collection of enciphering methods and examples, thanks to Trithemius, Selenus, Vigenère, and others, we might expect to see the same variety in practice—but here, too, we will be disappointed. In sharp contrast to the sophistication of this field of science on a theoretical level, the vast majority of enciphering methods that were actually practiced in the sixteenth and seventeenth centuries (and that go beyond the outdated monoalphabetic substitution) are of a type that is hardly mentioned in the handbooks: the homophonic cipher.[15]

Homophonic ciphers were designed in the fifteenth century by those who realized that frequency analysis made monoalphabetic codes easy to break and thus fairly useless. Their primary aim was to make the codebreaker's life hell, and they devised a method that was simple, easy to follow, and practical, and required only a one- or two-page key. These cipher keys—in their mature form—consisted of the following five categories:

1. Homophones—that is, three or four different characters that are assigned to each letter of the plaintext. More common letters are usually assigned more characters than less common ones. The purpose is to make frequency analysis impossible.
2. Special characters that are assigned for the most common double letters.
3. Special characters that stand for syllables.
4. Nullities—that is, characters that do not carry meaning—designed to confuse the codebreaker.
5. A nomenclator table—that is, a list of code words that stand for the most common conjunctions and prepositions, geographical names, and political players in the text. Because each of these is given a special sign or number, the system is more likely to resist the "probable-word attack," which looks for a word that is most probably found in a plaintext and tries to find its corresponding pattern in the ciphertext.

The use of two or three homophones for each letter results in an alphabet of approximately one hundred characters. There are usually no more than ten nullities, and the same number of letter pairs. Characters that stand for syllables usually number between 100 and 150, while a dictionary of code words could contain 300 items or more (although a very high number of code words makes enciphering more difficult) (see fig. 19). Homophonic ciphers were used in the early modern period in the diplomatic correspondence of Italy,[16] Spain,[17] France,[18] Germany,[19] and Hungary.[20]

This method prevailed until the end of the seventeenth century in both diplomatic and military correspondence. Progress within the system was due to the fact that the nomenclator dictionary was becoming longer and longer. By the time of Louis XIV, nomenclators of five hundred words were not at all uncommon. But this can be regarded as progressive only in one sense. When a cipher system (in which letters and combinations of letters are assigned to characters) develops into a code system (in which complete words are represented by certain characters), it becomes more secure. One disadvantage, however, is that both sender and addressee must own a rather thick dictionary that shows which character stands for which word. Security is not cheap, and the price you pay is a much longer key, or decreased user-friendliness, as we might put it today. At the same time, even this increased level of security is relative—given enough enciphered messages, the codebreaker may draw logical conclusions from the supposed content, a b c d e f g h i l m n o p q r s t u x z q z

Qua² que qui quo ⟨⟩ che come nuy vuy non per la mᵗᵃ V. la S. V.

Nihil importantes ñ ñ ∞ ∞ Λ Y ⤳ ⤳ pp ⧧ p⟩ ⟨⟩ φ̃ φ̃

ab	ce	eb	aʒ	ib	b	ob	⟩	vb	ι
ac	e	ec	a	ic	b	oc	p̄	vc	ī
ad	ʂ	ed	a	id	b	od	p̃	vd	ī
af	cʒ	ef	aa	if	b	of	⟩	vf	b
ag	c̄	eg	ā	ig	b	og	⟩	vg	ij
al	ca	el	ab	il	%	ol	⟩	vl	J
am	ce	em	ac	im	bʒ	om	⟩	vm	ιa
an	ci	en	ad	in	ba	on	⟩a	vn	ιe
ap	co	ep	ã	ip	be	op	⟩e	vp	ιo
ar	cu	er	a	ir	bi	or	⟩e	vr	ιo
as	es	es	a	is	bo	os	⟩i	vs	ι∞
at	et	et	āā	it	bo	ot	⟩o	vt	ιo
ax	eι	ex	ãã	ix	bu	ox	⟩o	vx	ju

Papa	3	Firentini	3⟩	S. Roberto da San		
Colegio di cardinali	3	Duca de Calabria	3u	Sever°	3	
Imperatore	3	Hippolita duchesse		Senesi	3	
Re di Franza	3	de Calabria	37	Bolognesi	3	
Re darayona	3	Duca de Modena	3	Duca Johanne	3	
Re de Ingalterra	3a	Marchese de Mantoa	3	Re Renato	3	
Duca de Bregogna	3e	Marchese de Mon-		Zenonesi	3	
Duca de Savoya	3i	ferra	3	Nave	3	
Duca de Milano	3o	Conte de Urbino	3	Galee	3	
Venetiani	3o	S. Alexandro	3	Duca de Bari	33	
				Filippo monsignore	33	

FIG. 19 | Homophonic cipher from Milan, 1448. At the top are the homophones assigned to the letters of the alphabet, then the characters that stand for conjunctions, followed by the nullities (*nihil importantes*), the signs that stand for the bigrams, and finally the dictionary of nomenclators, or code words. The characters in each of these five categories look somewhat different from one another, making coding—but also decoding—easier. From Aloys Meister, *Die Anfänge der modernen diplomatischen Geheimschrift* (Paderborn: Ferdinand Schöningh, 1902), 30.

the context of and the relationships between the code words. And if the codebreaker happens to obtain a copy of the relevant dictionary (if he steals it or, in the case of a historian, simply looks it up in the archives), he will have an easy job.

Otherwise, his task may become almost impossible. The story of the man in the iron mask illustrates this well. Voltaire was the first to speculate about this man's identity, and Alexandre Dumas elaborated on the story in his novel *The Vicomte of Bragelonne: Ten Years Later*. The story concerns a young man who was shipped under heavy security to the island of Sainte-Marguerite on the French Riviera in 1691, on the orders of Louis XIV. The prisoner, wearing an iron mask, was looked after by the governor of the prison himself. Nobody could meet him during his life, and not even at his death in 1703 was his identity made public. A number of theories have been advanced about the identity of this mysterious prisoner, some of them based on a letter that contains the most important piece of information—his name—in the form of a number group serving as a nomenclator—in other words, a code. The supposedly quite widespread nomenclator dictionary was never found, despite the best efforts of historian-cryptologists, so the debate over the prisoner's identity was never laid to rest.[21] Without the nomenclator key, a homophonic cipher can elude even the best codebreakers.

CRYPTOGRAPHY IN HUNGARY

In order to understand the context in which the Rohonc Codex was probably created, it is necessary briefly to review the background of early modern Hungary. The Hungarian territory was a particularly unstable, and consequently a very eventful, region between 1500 and 1700, one in which diplomatic, scientific, and religious secrets and attempts to decipher them played a crucial role. The states of east-central Europe became the battlefield for the war between Christianity and Islam, Catholicism and the Reformation, Western Christianity and Eastern Orthodoxy. Hungary, covering the whole of the Carpathian Basin, was seen by contemporaries as a rich and powerful country in the fifteenth and early sixteenth centuries. In 1526, however, it was first defeated by the Ottoman Turks and then, after the fall of its capital, Buda, in 1541, partially occupied by the Ottoman Empire. Subsequently, as one historian recently put it, Hungary became "a complicated set of lands caught up in an intricate network of alliances, belonging to and claimed by several ruling houses and dynasties."[22] As a result of a series of internal fights, the kingdom became divided into three parts: its central part remained occupied by the Ottoman sultan until the end of the seventeenth century; its western and northern regions continued their existence as the Kingdom of Hungary under the Habsburg kings but, owing to its geographical location, became a permanent battlefield between the Turkish and Christian armies; and the Principality of Transylvania in the southeast became a vassal state of the Ottoman Empire, with very limited independence.[23]

One peculiarity of the region is the multitude of languages used in everyday life. The use of Latin was a common feature of east-central Europe in early modern times, and Latin was the primary language used in diplomatic correspondence; German was widely used in the lands of the Habsburgs; the Pashas of Buda carried on their official correspondence with the Habsburg rulers in Hungarian, over the heads of the Hungarian dignitaries, while Hungarian nobility also preferred using the mother tongue in letters; Serbian and Bosnian spies sent their reports in Italian from Constantinople, or Ragusa (present-day Dubrovnik), to Vienna through Venice; and, of course, Turkish played an important role in and around the occupied territories.

Early modern Hungarian kings, princes of Transylvania, archbishops, and statesmen used the same type of homophonic ciphers as their western European colleagues when communicating with their envoys, generals, and fellow politicians. The Habsburg kings' dispatches to their regular ambassadors in Constantinople were also enciphered in that way, as were the letters of Ferenc II Rákóczi, the leader of the Hungarian uprising against the Habsburgs in 1703–11, to his own representative in the Ottoman Empire.[24]

A distinctive example of homophonic ciphers can be found in Archbishop Péter Pázmány's Italian correspondence.[25] Pázmány was not only an important prelate and theologian but also a cardinal and a crucial figure in the Hungarian Counter-Reformation. In the early 1630s, he was in conflict with Pope Urban VIII; thus it was of critical importance that he obtain firsthand information about the Italian political scene. C. H. Motmann, Pázmány's agent in Rome, sent him both encrypted and nonencrypted reports. Motmann used two-digit numbers (such as 24) for the letters of the alphabet—usually two such numbers (called homophones) for each letter—but vowels often were given three. He employed a combination of a consonant and a number (such as n7) for syllables, and he used three-digit numbers (such as 243) as nomenclators to represent political and geographical names. All of these characters matched the letters, syllables, and code words systematically, in an ascending alphabetical order (13, 15, 16 = A; 18, 19 = B; 20, 21= C; 22, 23 = D, etc.). This feature made the system practical for the people involved but also rendered it vulnerable. It also helps today's codebreaker historian, who has an easy job reading the code once he recognizes the logic behind the system.

Pázmány's cipher was deciphered by Hungarian historian Péter Tusor in cooperation with a retired Hungarian Signal Intelligence operative named Imre Máté. It illustrates the significance of diplomatic sources that still await study. Their content cannot be of minor importance, since they were enciphered; in fact, they might carry classified information that other contemporary sources do not contain. This cipher is also worth mentioning because its solution was the product of cooperation between a historian and a mathematician-codebreaker. Because of their complexity, homophonic ciphers typically require such joint efforts. While identifying the letter and syllable signs requires outstanding mathematical, linguistic, statistical, and logical reasoning skills, the nomenclators that stand for geographical and political figures can be identified only with extensive knowledge of the historical context.

Ciphers were used not only in formal diplomacy but also by the people involved in one of the numerous anti-Habsburg conspiracies. The political situation in tripartite Hungary, partly under Habsburg reign, was fairly delicate and marked by conflict with the ruling dynasty. Historians today regard the relationship between the Kingdom of Hungary and the Habsburgs as highly complex,[26] but contemporaries, both peasants and nobles, often had the impression that sixteenth- and seventeenth-century Hungary was merely a buffer zone between Austria and the Ottomans, exploited and oppressed by its foreign kings. As a consequence, and often with the intention of regaining some of the kingdom's lost autonomy, conflict and discontent grew into anti-Habsburg conspiracies, uprisings, and, once, even a war of independence. The best known of these conflicts are the conspiracy led by the Palatine Ferenc Wesselényi in the 1660s, the uprising of Imre Thököly in the late 1670s and early 1680s, and, finally, Ferenc II Rákóczi's War of Independence (1703–11). It was critically important that these politicians and military leaders maintain secrecy in communications with their generals and foreign allies (including the French king and the Ottoman emperor).

More than seventy different cipher keys have survived from the environment of Ferenc II Rákóczi, each of them allegedly used to encipher several letters. These methods were complex and fairly modern by the standards of the age.[27]

A close reading of these cipher keys reveals strange secrets. Comparing, for example, Rákóczi's two most elaborate keys, curious details emerge. One of the keys is a typical homophonic cipher key, with a long list of nomenclators naming geographical territories and political figures in French, enabling sophisticated diplomatic and military communication. This key was used to encode correspondence between Rákóczi and his primary ally, the French king Louis XIV. Because of its importance, it survived in many other, less artistically elaborate copies. The other elaborate cipher key looks almost the same, and its language is also French, but its nomenclators name only two geographical units, Krakow and Warsaw, and not a single politician. It also contains certain words that are missing not only from the other complex cipher key but also from most, if not all, of the keys that survived: *abandonne*, *adorable*, *chagrin*, *jaloux*, *solicitude*, *sentimans*, and *souvenir*. There is no clue as to the recipient of the letters encrypted with this code, but it must have been someone close to Prince Rákóczi's heart, someone as important as the French king. There is a love poem in French on the back of a copy of this second key, with certain typical mistakes that indicate that the prince himself must have written it to his close friend and lover, the Polish stateswoman and patron of the arts Elżbieta Helena Sieniawska (1669–1729), with whom he was corresponding in French. We have every reason to believe that this key was never used for diplomatic messages; instead, it served to protect the privacy, not to say the love affair, of the prince.[28]

In addition to the communiqués of official diplomacy and resistance, the personal correspondence of nobility was often enciphered—particularly when the content of the message was of a political nature. Members of the Hungarian nobility, especially those who played important roles in the internal politics of the Kingdom of Hungary or Transylvania, or both, usually wrote to one another in Hungarian plaintext. However, it is not unusual to see certain words, expressions, and parts of sentences written in homophonic ciphers. What they chose to encipher reveals the areas in which they had to be cautious. For example, Transylvanian politicians sometimes had to pay more attention to the Turkish threat than to the Habsburg reaction: they carefully enciphered insults directed at the Turks (the word "pagan" was the usual derogatory term for them, and it was always enciphered), but left insults regarding Habsburg officials in readable plain text.[29]

Ciphers also played a crucial role in espionage, of course. Secret spy networks were embedded in the territories between the Habsburg and Ottoman Empires—that is, the Kingdom of Hungary and Transylvania. When a new agent was admitted to the system of correspondents, the encoding system was established. The scope of espionage extended well beyond diplomacy; it involved all strata of society. Local doctors, merchants, soldiers, and ambassadors' interpreters (that is, everyday people far from the diplomatic hierarchy) risked their lives collecting and transmitting information. Effectively disguised by their quotidian professions, these people caused little suspicion in their comings and goings. Spies in the region were most often of southern Slavic origin; in other cases they were Armenians, renegades, or Sephardi doctors. Often, families provided several generations of agents. Examining the source material, the historian not only can identify their names but also can describe their cultural backgrounds, what they were paid for their services, the risks they took and the nature of the information they carried, and their fears. These things can be measured by how seriously the Habsburg court took these letters and by the speed with which the enciphered letters were decrypted. Some of the spies were double agents working simultaneously for the Ottoman and Habsburg courts, perhaps using both Islamic and Christian encrypting methods.[30]

People who occupied considerably less dangerous positions, among them university students, craftsmen, and even the first person to write poetry in Hungarian, also used enciphering techniques. Late fifteenth-century university manuals, for example, contained sophisticated secret alphabets scattered among scientific texts on astronomy, astrology, and medicine. Interestingly, these secret alphabets remained on a somewhat primitive level and followed the outdated monoalphabetic cryptographic system.[31] One example is the magical *Liber runarum*, the Book of the Runes, several copies of which survive in students' handbooks. This is a short Latin text on manipulating planetary spirits for all kinds of benign and malign aims with the help of talismans. Here, the spirits' names are to be engraved on metal plates in a runic alphabet. This is quite typical in university students' handbooks: the alphabets of secret letters in the magic texts were not applied primarily as a cryptographic tool but rather to be engraved on talismans to designate spirits. These secret alphabets were too outdated to be real ciphers, and it is not their lack of cryptographic sophistication that interests us here. Easy to break, they do not protect a text's content; instead, they serve to attract readers' attention and make the text more appealing to them, and they serve to communicate with planetary spirits. It is through the use of such signs that the user of magic might get closer to the spiritual realms.[32]

In the scientific and artisanal tradition, reasons for secrecy were different, but the need was equally pressing. The Renaissance engineer Giovanni Fontana (ca. 1395–ca. 1455) applied simple substitution ciphers in order to describe complex technical constructions and military machines.[33] Galileo (1564–1642) encoded his scientific conjectures in anagrams in an age when copyrights and patents were in embryonic form and could not attest to who discovered something first.[34] Robert Boyle (1627–1691) enciphered his alchemical findings in order to protect the results of his experiments.[35] In Hungarian history, a mid-sixteenth-century source, a private diary of 120 pages, is relevant here. It belonged to Johannes Cementes of Kolozsvár (from Cluj-Napoca in present-day Romania), supervisor of the mint and a master assayer. Cementes kept a record of events in both his private and professional life from 1530 to 1586. In the Hungarian text, he applied simple ciphers to encode details of his mining, minting, apothecary, and alchemical knowledge, most often single words but including the title itself.[36] The method he used is not nearly as well developed as the homophonic systems applied in contemporary diplomacy and the private correspondence of the aristocrats. Like the late medieval students in their handbooks, Cementes applied the simple and vulnerable monoalphabetic substitution system.

The Hungarian poet, soldier, and seducer Bálint Balassi (1554–1594) also played a role in the history of ciphers. Balassi did not lead the conventional life of an introverted poet. A number of towns pursued him over discipline issues because of his scandalous and pugnacious character. He was embroiled in an endless inheritance dispute with his relatives, and he had a never-ending series of love affairs to add to his problems. During his eventful and difficult life, he wrote many letters about private and financial secrets in which he usually applied partial enciphering. He tried to hide the meaning of particular words—sometimes names, sometimes vulgar sexual expressions—in an otherwise readable text. This method was even simpler than the ciphers used by Cementes or the university students. Not only did Balassi use a monoalphabetic cipher, but he chose a very simple, even primitive version: the letters in the first half of the alphabet corresponded to the letters of the second half of the alphabet. Instead of A, he wrote M; instead of B, he wrote N; instead of C, he wrote O, and so on, following an easily breakable system. The best codebreakers at the Habsburg court were not likely to be interested in Balassi's money issues and romantic trysts, but his cipher system was secretive enough to keep the prying eyes of his messengers out of his personal business.[37]

Sexual secrets and other private matters were enciphered in other people's diaries as well. The Transylvanian politician Gábor Haller (1614–1663) left a detailed and straightforward personal diary to posterity; however, when he wrote about the quantity of alcohol he drank, and the consequences of his drinking, he enciphered key words. Haller also recorded his sexual desires and a few political secrets in ciphers, using two different yet very simple systems. One of them was a version of the famous pigpen cipher later used by the Freemasons; the other was an easily recognizable letter-transposition method. But, wanting to make sure that his secrets would be easily revealed, he copied the cipher keys at the end of his diary![38] Another Transylvanian nobleman, Zsigmond Szaniszló (ca. 1655–ca. 1721), also used ciphers in his private diary. Usually detailing financial secrets, on one occasion he recorded that his wife—heavily pregnant, as we learn in another part of the diary— spent the night in their guest's bedroom. Had the author wished to hide this information, he could have chosen a more complex cipher than the one he used, which only substituted the vowels and left the original consonants in place. In his case, shame seems to be the motivation behind secrecy.[39]

Such easily decodable ciphers could be found in others' writings as well. The influential politician Palatine István Illésházy (1541–1609) used them in 1605, when he was in exile, in a private letter to his wife, Katalin Pálffy. The otherwise unknown author Miklós Nagy also used a simple cipher in a few paragraphs of the so-called Kuun Codex. These two ciphers left the consonants of the plaintext in place and encoded only the vowels, even then according to an easily recognizable system. The relevance of these sources lies not in the sophistication of their cryptographic system but in the fact that they were used by common people for their private purposes.[40]

These examples demonstrate that the use of ciphers was not restricted to state affairs in early modern Hungary; literate people of various social strata used them in their everyday lives. In fact, this widespread use of ciphers has led one historian of the period to argue that Hungary was an exceptionally rich territory for ciphers.[41] To what extent this conclusion is correct, in comparison to other countries in the region (such as Poland, Bohemia, or the German principalities), is an open question that systematic research has yet to confirm.

Over the course of three hundred years, roughly from 1400 to 1700, scribes in state diplomacy gradually replaced monoalphabetic systems with more complex homophonic ciphers. By the end of the period, not only letters of the alphabet but also syllables had separate signs or numbers, while nullities and code words were used more and more extensively. In a parallel development, simpler encrypting techniques were introduced into various social classes and trades. The further we stray from the diplomatic world, the more dominant monoalphabetic ciphers become. This was probably due to the lack of professional codebreakers at the lower social strata, the less vital nature of the secrets to be protected, and, not least, the relative complexity of homophonic ciphers.

Various types of everyday users became part of the history of cryptography. The widespread use of ciphers and their growing popularity must have been related to the fact that in tripartite Hungary, partly occupied by the Turks, partly living under Austrian authority, and partly balancing between the two in Transylvania, a large portion of the population lived in a frontier zone and participated (or was forced to participate) in the network of information flow as possessors and transmitters of secrets.

This was the cryptographic background of the age in which the Rohonc Codex was written.

CHAPTER 6

DECRYPTION

CODEBREAKING THEN . . .

The science of cryptology has long since turned to more serious problems than breaking early modern ciphers. We are far beyond the World War II problem of the Enigma code. Today's focus is on devising secure methods of sharing a key and on computerized encryption techniques that laypeople can hardly comprehend. Given today's enormously complex standards and algorithms, a homophonic cipher from the seventeenth century should be easy to break. But is it? If there are too many nullities, if symbols are assigned not only to letters but also to syllables, if there are no obvious word or even letter boundaries, if there is an extensive table of nomenclators, if the language of the plaintext is not known—if even one or two of these conditions obtain, let alone all of them—then even an outdated cipher system may be extremely challenging.

Knowing the historical context is helpful, and so is correctly identifying the plaintext language and having an adequate knowledge of mathematics, a patient attitude toward trials, and a good deal of ingenuity. Carelessness in encryption may also make the job of the codebreaker easier. There might be accidental spaces left between words, or plaintext fragments in the encrypted text (thus revealing the original language). And even though, in theory, there are three or four different signs in a homophonic cipher that should be used to substitute for the same letter of the plaintext, the actual user may not always exploit this variation.

In the absence of such carelessness, a ciphertext may remain unbroken for decades or even centuries, even if, theoretically, it is not unbreakable. Beatrix Potter (1866–1943), author of *The Tale of Peter Rabbit* and other children's books, wrote an encoded diary between 1881 and 1897, which, though not at all beyond the capacities of a well-prepared codebreaker, was not broken until 1958. Even editors of modern editions of originally enciphered letters and other texts rarely *decipher* them; instead, they merely *publish* the letters (which typically were decoded by a contemporary of the writer, most often the addressee). When a code is not cracked within a few months of its creation, it is most often unbroken to this day. The cryptologically informed historian is thus faced with a challenge when she wants to reconstruct sources or segments of letters that are not available to readers today.

But what resources can the historian call upon when trying to break a cipher? She should begin by studying some manuals on codebreaking so as to learn the art of cryptanalysis.[1] Handbooks and manuals there are, and there were in the past as well. Should the text that she wants to decipher be from the seventeenth century, then the adventurous historian had better learn about the encryption and decryption approaches used at the time, in addition to studying modern handbooks. This is how I approached the challenge of tackling the Rohonc Codex.

Proceeding chronologically, I started with the major early modern codebreaking summaries. In the sixteenth and seventeenth centuries—just as in later history—these were fairly rare. The handbooks that were available to the general population made sure not to reveal the latest secret methods in cryptology. State-of-the-art handbooks were accessible only to social and political elites. This remained the case until quite recently, though it may surprise us to learn that even in 1967 cryptologist-historians complained about the thoroughgoing lack of available information about the codes and ciphers of World War II.[2]

Not that historical codebreaking handbooks are always helpful. The author of a 1641 handbook, Antonio Maria Cospi, secretary to the Prince of Tuscany, admits that his competence does not extend beyond the sphere of simple monoalphabetic ciphers, and he steers clear of detailing the homo-phonic, or, in his words, complex ones (*chiffre composée*). Declaring these to be unbreakable, he offers detailed methods for cracking monoalphabetic ciphers, explains how vowels can be identified, and lists the most frequent syllables in French, Spanish, and Latin.

Luckily, other writers were braver. The French lawyer and amateur mathematician François Viète (1540–1603) is remembered as the "father of algebra." Less well known is the fact that he regularly performed cryptanalysis for Henry IV, the French king, who was haunted by his rivalry with the Habsburg rulers of Spain. In a secret letter to King Henry's minister, the Prince of Sully, Viète describes his "unfailing" codebreaking method, the so-called infallible rule, through which he identifies vowels by analyzing the frequency of the combination of double and triple characters in the cipher-text.[3] Since all languages have fewer vowels than consonants, and since their location in a word clearly identifies the structure of that word, vowel analysis is a crucial part of all cryptanalysis. Viète's study begins with the typical code signs of the rival Spanish court. He points out that these Spanish codes usually use three or four symbols for each letter, one or two for each syllable, several series of characters for the most frequent words and proper names, and special signs for double letters. He discusses frequency analysis, which was not really a novelty at the time, and then analyzes the triads and dyads of a text, which enables the hidden vowels to be discovered.

This method is indeed effective in breaking monoalphabetic ciphers, in which every character stands for a given letter. Homophonic ciphers, as we have seen, are a great deal more complicated, particularly when they also encipher syllables. Contrary to what Viète suggests, his method has virtually no hope of breaking a homophonic cipher unless the text is strongly formalized and has an identifiable address, greeting, signature, date, and other components. Applying the "infallible rule" will be especially difficult if nomenclators are used extensively and are not easily distinguishable from the symbols of letters and syllables. We have every right to suppose that Viète, the talented mathematician and experienced codebreaker, could successfully find vowels in homophonic ciphertexts, too, and thus break a given code—and perhaps he could. But he is careful, in describing his infallible rule, not to give too much away or equip his readers with really useful tools, presumably wanting to protect his methods from King Henry's enemies, lest his letter fall into the wrong hands.

Fortunately, we have more informative sources. The most didactic cryptological handbook of the early modern period is by an anonymous (supposedly Spanish) author. "The Art of Deciphering" ("L'art de deschiffrer") is a 136-page manuscript that survived in French from the seventeenth century.[4] Though it does not offer any "infallible" solutions, it does provide a number of observations, maxims, rules, and recipes that can help identify the language of the plaintext, identify which signs stand for letters and which for syllables, recognize nullities, analyze letter frequency, and so on. Whoever learns these skills will be able to decode a cipher created in this era. The author distinguishes between simple (monoalphabetic) and complex (in other words, homophonic) ciphers, gives a detailed analysis of both, and introduces a case study for each, wherein he breaks a ciphertext step by step, applying his own maxims and recipes. In terms of theory, "The Art of Deciphering" is not much more valuable than Viète's secret letter. What distinguishes this manuscript is that it is longer, more detailed, and much more practical, providing a number of specific examples that illustrate the application of its principles.

. . . AND NOW

Why bother with a pen-and-paper frequency analysis, extensive search for vowels, and tedious note keeping when we have superfast computers, the calculating capacities of which exceed those of humans by several orders of magnitude? Are they not capable of breaking codes a hundred times more difficult than the ciphers of yore?

The short answer is that computers can perform only the tasks that we set them (albeit at a much greater speed than we can). Even a child can write software that counts the frequency of the characters in a text that we have fed into a computer. It is the feeding of the computer that can be problematic. Imagine a text that you possess only in manuscript. Imagine that your text contains symbols unknown to you. In this case you must start by assigning a number to each and every symbol in order to be able to digitize the script. Sound easy? Suppose that a creature from Mars comes to earth with a superintelligent computer. The Martian wants to decode the English language but has only a modern handwritten letter as a source. Her first task would be to code the unknown symbols (the letters of our alphabet) into manageable characters. But how would a Martian know that the letters *n* and *u*, which look very similar in a handwritten text, are actually two distinct characters? How could she know that the letter *m* is not a double *n* or a treble *i*, or, for that matter, that *M* and *m* are the same letter? Our Martian must make quite a few decisions before setting her superintelligent computer to work.

We humans confront the same scenario when we try to decipher a text. Computers are incredibly fast at testing a hypothesis or calculating statistics; they can tell you in a flash which characters appear next to which, how many times, and so on. But we still need a human who is capable of digitizing unknown characters and formulating questions and hypotheses. Captain Parker Hitt, the author of the first cryptography handbook in the United States, believed that such a person needs four essential things: perseverance, careful analytical methods, intuition, and luck. Computers can provide only the first two of these.[5]

In short, the most important steps of the deciphering procedure require human intervention, and it can be extremely demanding and nerve-racking work. It is not surprising that William Friedman, one of the most famous codebreakers of all time, suggests at the beginning of *Military Cryptanalysis*, his multivolume manual for codebreakers, that when a person becomes exhausted or overwrought while working on a cipher, he should take a break—by playing a game, for example.[6]

Though not set in a military environment, John Chadwick's linguistic research is instructive. This is how Chadwick, who took part in breaking the ancient Greek writing Linear B, sees the process of breaking codes and mysterious ancient texts:

> Cryptography is a science of deduction and controlled experiment; hypotheses are formed, tested and often discarded. But the residue which passes the test grows and grows until finally there comes a point when the experimenter feels solid ground beneath his feet: his hypotheses cohere, and fragments of sense emerge from their camouflage. The code "breaks." Perhaps this is best defined as the point when the likely leads appear faster than they can be followed up. It is like the initiation of a chain reaction in atomic physics; once the critical threshold is passed, the reaction propagates itself. Only in the simplest experiments or codes does it complete itself with

explosive violence. In the more difficult cases there is much work still to be done, and the small areas of sense, though sure proof of the break, remain for a while isolated; only gradually does the picture become filled out.[7]

First and foremost, cryptanalysis requires a human mind in which this process can take place. Such a mind, however ingenious, will need to be armed with a series of cryptographic methods. These can be learned through formal training and from handbooks of cryptography.[8] A person is said to have successfully learned these methods once she has actually deciphered a ciphertext.

In general, a cipher can be broken only given an adequate amount of ciphertext. Very few encoding methods have withstood concerted cryptanalytic efforts, provided the ciphertext is of sufficient length. As Edgar Allan Poe, an expert on the subject, put it, "human ingenuity cannot concoct a cipher which human ingenuity cannot resolve."[9]

Given an adequate amount of ciphertext, the process of deciphering consists of these steps:

1. Identifying the language of the plaintext.
2. Identifying the method of encryption.
3. Reconstructing the actual key.
4. Reconstructing the plaintext.[10]

These steps normally follow this order, although the second sometimes precedes the first. In any case, if neither the plaintext language nor the encryption method can be identified, the process cannot move on to steps 3 and 4. Let us take the steps in order.

1. At the beginning of his classic textbook series, William Friedman refutes the widespread notion that a code or cipher can be broken without knowing the language of the plaintext (this means both identifying the language and having some fluency in it):

The writer has seen in print statements that "during the World War ... [decryption experts] decoded messages in Japanese and Russian without knowing a word of either language." The extent to which such statements are exaggerated will soon become obvious to the student. Of course, there are occasional instances in which a mere clerk with quite limited experience may be able to "solve" a message in an extremely simple system in a language of which he has no knowledge at all; but such a "solution" calls for nothing more arduous than the ability to recognize pronounceable combinations of vowels and consonants—an ability that hardly deserves to be rated as "cryptanalytic" in any real sense. To say that it is possible to solve a cryptogram in a foreign language "without knowing a word of that language" is not quite the same as to say that it is possible to do so with only a slight knowledge of the language; and it may be stated without cavil that the better the cryptanalyst's knowledge of the language, the greater are the chances for his success and, in any case, the easier is his work.[11]

The British military pamphlet *Enemy Codes and Their Solution* emphasizes on page 1 that a cryptologist must be thoroughly familiar with the vocabulary, grammar, syntax, and idioms of the underlying language.[12] This is something even medieval Arabic tracts warned about.

Identifying the plaintext language may be possible by traffic analysis—that is, analyzing the communication of the enemy. In the case of early modern secret diplomatic correspondence, this may be done simply by reading the address and the signature, since such letters were only partially enciphered, with the encrypted and plaintexts mixed, to the delight of the cryptologist.

In the usual war situation, when it is clear who the enemy is and what language it uses, this is an easy step. But even this can sometimes cause surprises. In World War II, before the Americans became involved, the Germans wrote their messages to their ambassador in Washington in English rather than German, which the nosy Americans quickly realized. The Japanese, however, were baffled by the messages that the Americans sent one another, because they used the language of the Navajo Indians. The use of a language with such an irregular structure, unknown to the enemy and not described in any textbook, was ingenious and proved surprisingly successful. Does the failure to identify the language of the plaintext create a similarly insurmountable challenge in the case of the Rohonc Codex? We shall return to this question later.

2. In an ideal situation, the next step is to identify the type of encryption method applied. This, perhaps the most critical part of the process,

determines the outcome of the undertaking, and it requires the largest amount of work. If possible, the decoder should determine what encryption methods the sender has used previously, a task involving research. The job of the historian in such cases is to find historical analogies. A source from sixteenth-century Spanish diplomacy is likely to be written in a homophonic code that includes extensive nomenclators, whereas a thirteenth-century letter from the pope's chancellery is more likely to be written in a simple monoalphabetic code. The exact nature of the homophonic or monoalphabetic system should be identified in the next step.

When this type of analogical reasoning does not produce results, or when it is too problematic to carry out, one can try examining the inner structure of the cipher. The statistical results, including a thorough frequency analysis, the basic appearance of the text, examining entropy and other tests, often reveal whether the code is mono- or polyalphabetic, coded letter by letter or syllable by syllable. One Dutch matrix code was broken when the decipherer noticed a simple feature of the document, namely, that certain characters were placed slightly lower on the line than others. It was rightly supposed that these characters might have been written into the string of characters through a matrix. Putting these characters together created meaningful word fragments. Having reconstructed and carefully rotated the matrix, a turning grille, the decoder was able to reveal the encrypted German text.[13]

Sometimes nothing is known about the background of the sender or recipient, and even the text analysis fails us. The process typically applied in this situation requires an unusual amount of effort. Every possible encryption method must be tested one by one. Friedman's book (like al-Kindi's ninth-century treatise) contains a gigantic analytical chart that lists every encryption method imaginable. Testing the methods listed there one by one, the codebreaker must use the process of elimination until only one is left.[14] Few activities are more tiresome.

3. Reconstructing the key of the cipher paves the way to the complete solution. Although this necessary step may be performed only partially, it is worth finishing before rushing on to the final step of total text reconstruction, since it speeds up that final step considerably. This stage consists of producing a complete cipher alphabet and the entire list of homophones and nomenclators. In the case of a polyalphabetic system, it means defining the sequence of the code alphabets; in the case of a grille, it means creating the matrix.

4. At last we arrive at the most productive and satisfying stage, the reconstruction of the plaintext. Needless to say, steps 3 and 4 cannot always be separated completely: sometimes a few deciphered symbols help us recognize a certain word, the characters of which are then put into the cipher alphabet under construction. And thus one moves on, alternatively working on steps 3 and 4.

These, then, are the stages of the procedure. But what actual tools are applied when following these steps? Traffic analysis and, in the case of historical codes, the study of the historical background always lead to important insights. For this reason, I tried to learn as much about the history of the Rohonc Codex as possible. I studied its watermark, I looked for its traces in libraries, I collected historical analogues so as to narrow down the number of likely encryption methods.

Having researched the historical context, the codebreaker starts the statistical examinations of the text. The frequency analysis of the characters—explained in the previous chapter—is usually the first method. As we have seen, cryptology handbooks contain extensive charts of the typical frequency patterns of certain languages. The most or least frequently used letters of a language are usually revealing—for the codebreaker, both categories are equally important. These features, however, are not always reliable. Although *e* is the most frequent character in both English and Hungarian, there are quite a few English and Hungarian ciphertexts that simply do not follow a pattern in which *e* is used more frequently than any other letter. Letter combinations have their typical frequencies, just as letters do, and tables of these can also be found in particular handbooks. Frequency analysis therefore must determine which two characters often stand together. Similarly interesting are the most frequent combinations of three or four letters.

Repetitive patterns are a source of great joy for the codebreaker, who suspects that they either hide similar texts or at least are similarly encrypted. The number of characters between two such repeated patterns must be counted. If the result is always an even number, this might mean that we are dealing with a two-character code, which means that in the continuous string of multiple-digit numerals, only the two-digit numerals carry *meaning*. The distance between the repetitions may also indicate how frequently a given code alphabet is repeated in a polyalphabetic code.

Codebreakers are interested in differentiating vowels and consonants. Curiously enough, some characters show themselves to be either vowels or consonants even at a very early stage, long before their meaning is discovered. These two types of letters behave in different ways. Vowels are fewer, and they typically stand next to many different characters (consonants). Consonants are much pickier about their neighbors. The most frequent character in a text is almost always a vowel, and the least frequent is a consonant. The characters usually standing next to the most frequent letters (which are vowels) are consonants. There are several methods based on these simple facts that can be used to reveal vowels, the two easiest of which are the pen-and-paper method of trilateral frequency analysis,[15] which can be applied even with very brief texts, and a Russian method called Sukhotin's algorithm.[16] Adventurous American codebreakers, using a mere half-sentence script, discovered that the latter method identifies the vowels in such exotic languages as Gruzian, Gaelic, Croatian, Hebrew, and even Hungarian.[17] Admittedly, this algorithm occasionally mistakenly identifies a consonant as a vowel, but only after correctly identifying all the real vowels.

Since I did not come across any attempts to use Sukhotin's algorithm on a historical homophonic cipher, I tried it myself. The result exceeded my expectations: Sukhotin's algorithm successfully reveals the vowels even in a homophonic system. While these ciphers can disguise vowels as highly frequent letters (hindering frequency analysis), they cannot hide them as letters that are in contact with a large number of other letters. This is why Sukhotin's algorithm manages to find them. The technical details of both trilateral frequency analysis and Sukhotin's algorithm can be learned in ten minutes.

Similar to the charts of letters and letter combinations, but even more useful, are the so-called word-pattern charts of cryptology handbooks. During the course of codebreaking, we are often faced with a situation where we know the pattern of a word. We know which letters are identical and which are different, and we want to know which words of the given language fit this particular pattern. Obviously, a word with an ABAB structure is not of much use, but a word with a more exotic structure, like *attack* or *division*, in a military text, or *Barabbas* in a biblical text, will often prove to be a point of entry into the cipher. Beware, however: it is not enough to identify a word with the structure of *attack* (ABBACD), since that word might also be *effect* or another word with a similar structure. This is when frequency analysis comes into play.

The same method in reverse is called the probable-word method. For example, we might be certain that the encrypted text does contain a particular word, such as *enemy* or even *Barabbas*, so we start looking for words in the character flow that have the given structure of this word. The importance of this method is not to be underestimated. If it is successful, it might offer a basis for solving the complete code. The probable-word method may be applied not only to character strings with a typical structure but also to those that can be found in typical places in the ciphertext. In military messages, the first and last words and the first and last sentence parts are especially important. Since the structure of military orders is quite formal, we rightly expect them to contain at the beginning some kind of an address that states the topic of the message, a classification code, or a reference to other messages, and a signature or another type of identification at the end of the text. Such predictable features weaken the security of an encryption method, yet they are hard to avoid. The probable-word method may also work for a civilian cipher. A secret diary is likely to contain the name of the author or a date on the first page. And this brings us to the matter of numbers, another weakness of ciphers that can be used as possible entry points.

The encrypted diary of the nineteenth-century Italian Egyptologist Simeone Levi was broken on the basis of the numbers it contained. Numbers tend to occur in clusters, often consecutively. Or the same number will be repeated several times. As coordinates, dates, quantities, and years are an integral part of military documents and can often be found in nonmilitary texts as well, we may assume that the cipher we are studying will also contain them.[18]

The analytical methods discussed here are the easiest to apply to monoalphabetic ciphers. But cryptanalysis often consists basically of reducing a complex cipher to the level of a monoalphabetic system. Therefore, these methods often play a role in the final stage of cryptanalysis, if not earlier. There is one method, however, that has been specifically designed to attack a homophonic system, and it aims at finding an entry point to the cipher via analyzing the "isologues." Isologues are cipher sections that look different but have the same plaintext behind them. If, for example, a code key assigns several different characters to each letter, or if it assigns characters to both syllables and single letters, the result may be a ciphertext where the same word of the plaintext is encrypted in two different ways. With a little luck, one may see the same word encrypted several times, with different characters, in which the same letters are encrypted sometimes with identical and sometimes with different characters. For example:

Plaintext: in dis p en sa b le
Ciphertext: 44. 31. 22. 76. 96. 12. 71.
Plaintext: in d i s pe n s a b le
Ciphertext: 44. 7. 3. 2. 65. 16. 2. 5. 12. 71.

Such isologues appear when we look for repetitive patterns in the ciphertext. In this case, 44, 12, and 71 will appear in both cases, and the codebreaker might suppose that 31. 22. 76. 96 corresponds to 7. 3. 2. 65. 16. 2. 5. Recognizing the isologues that encrypt the same text in different ways, we may be able to assign a syllable sign to two character signs, or homophones, that belong to the same plaintext. This is a tiresome yet ultimately rewarding procedure.

There are obviously a number of additional cryptological methods, some of which require rather complex mathematical skills.[19] However, early modern ciphers and the Rohonc Codex do not require such complex tools, so there is no need to elaborate them here.

Finally, one misconception concerning cryptanalysis should be clarified. The fact that someone may spend long days looking at a code or cipher, trying one method after another without success but putting on a brave face, does not mean that she is just wasting her time while quietly going nuts. Cryptanalysis involves such futile-looking activities as transcribing the complete text by hand, for example, because this may help us recognize a pattern or feature otherwise invisible to the eye. Breaking a difficult but solvable cipher may easily require a thousand hours of work. Convert that into workdays!

IS THE ROHONC LANGUAGE A CIPHER?

Remember that modern cryptology handbooks suggest identifying the language of the plaintext first, followed by the second step, which is identifying the encryption method. I attempted the first objective by trying to discover the place of origin and historical context of the Rohonc Codex. Still, I have not been able to radically limit the number of possible languages. Since we have no clue about the author or his readers, and no other source has emerged that was encrypted with the key of the codex, traffic analysis could not be done either.

Let us focus, then, on step 2. William Friedman suggests that we apply the exclusion procedure in the absence of a better method. This entails looking at every possible cipher and determining whether one of them could have been used to create our ciphertext. By process of elimination, we draw the circle tighter and get closer to discovering the type of cipher actually used. This is a quite promising avenue.

It is clear enough that the Rohonc Codex was not encrypted using a simple monoalphabetic cipher. There are far too many characters, the repetitions are too frequent, and, to top it all off, the letter-frequency analysis reveals an alphabet that does not resemble any of the natural languages. Nor do vowel-consonant tests succeed in grouping the symbols into these two types. Thus I excluded monoalphabetic systems, and for similar reasons I ruled out the *simple* homophonic system as well. At this point, I still thought it made sense to consider complex homophonic systems, including nomenclators.

But how to make the number of options smaller? The polyalphabetic methods that use several code alphabets can also be excluded, for the entropy of the text is rather low. Apparently, the codex does not use a series of different cipher alphabets, because such a polyalphabetic method would be able to hide the obvious patterns. The codex is jam-packed with frequent and often immediate repetitions and with similarly structured passages, and a great many paragraphs begin in the same way. Some paragraphs are even repeated completely. Given these facts, a polyalphabetic cipher would be impossible.

As a next step, it is worth checking the methods that Gustavus Selenus lists at the beginning of his handbook. Selenus (or, more precisely, Augustus the Younger, Duke of Brunswick-Lüneburg, who wrote under the pseudonym Selenus) presents lengthy texts of innocent content in which only certain characters play a role, making up the hidden message character by character. These characters may be the first letter of every word, or every fifth symbol, or even—and why not?—just one character per page. Although the Rohonc Codex, as we have seen, does not contain Latin letters but symbols, many of these symbols closely resemble Latin letters. And if we mirror the pages of the book, even more of them look like Latin letters. But, try as I might, I have not been able to devise a system with which to distinguish between relevant characters and those that should be ignored. One important reason for this is that several letters of the Latin alphabet do not appear in the codex at all, mirrored or not, which makes it unlikely that a regular, meaningful text could be built from the few Latin-like letters that are actually in the text. Furthermore, this method would fail to explain the repetitions in the codex. We cannot exclude the possibility, of course, that every, let's say, sixth character plays a role in the text, and that these would make up a character string in a monoalphabetic or homophonic cipher that could then be deciphered with the help of the usual tests. It would take quite some time to conduct every test for every first, second, third letter, etc. The other factor that makes this theory less than likely is that the codex as a book looks somewhat worn. It must have been handled quite frequently, probably carried in a pocket, and we must assume that its readers were able to read it easily and often, without having to carry out a series of complicated decrypting procedures each time.

It would be more promising to try to reconstruct a complex homophonic key that would take into consideration all of the characters of the code text. A strong argument in favor of the homophonic system is that the script gives the impression that some of its characters stand for letters, while some others seem to represent concepts or word parts. It also suggests that some redundant symbols carry no meaning—that is, that they are nullities.

If these suppositions are correct, and the Rohonc Codex was made using a complex homophonic cipher, then we should find in it a large number of letter and syllable signs and a few word signs. It is, of course, hard to identify which symbols stand for letters and which for words or syllables. I tried to make a preliminary distinction between certain characters that I assumed to be signs for letters and those that I assumed stand for longer linguistic units, like syllables or words. Then I ran letter-frequency and vowel tests on what I judged to be letter strings. The only thing I concluded was that the sections I examined were probably not made up of vowels and consonants.

Then I systematically collected the symbols that directly precede and follow the symbols associated with the names of Christ and Pilate, hoping that they would make up a typical preposition, function verb, or suffix.

I tried probable-word analysis, too, assuming that the repetitive elements of the text could be the typical phrases of a religious text ("Lord," "pray for us," "have mercy on us," etc.). Comparing the script to the prayer books, breviaries, and Books of Hours of the period, I had to conclude that the number of repetitions in the codex is no higher than what we would expect to find in such books. And Gospel extracts, psalms, and prayers that, on the basis of the pictures, we would expect to appear in the Rohonc Codex are often included in breviaries and Books of Hours.

Part of my probable-word analysis rested on the assumption that the conversation between Christ and Pilate would resemble the relevant biblical text, because, say, the codex is a Bible commentary. If this is true, then the text surrounding the picture depicting Christ and Pilate should contain the words "Barabbas" and "Jews," as in the Gospel accounts. Between the signs for the names of Christ and Pilate I looked for signs that could mean "reply," "said," "to him," and so on. Yet I was not been able to find the word structures for these words in any known language.

Next, I looked for isologues, strings of characters that are repeated not identically but with a slight difference each time. As we have seen, isologues can be used in reconstructing a homophonic system. To my surprise, I found that when longer sections, even complete pages, are repeated, they are repeated with almost the same characters, something that is highly unlikely when the encryption is based on a code chart offering alternative possibilities. It is more probable that the second occurrence of such sections in the codex was either copied directly from the first, or that the cipher key contained only one equivalent for each letter. However, the fact that the INRI inscription occurs in two different versions in the images contradicts this theory.

Because of the behavior and diversity of the characters, I had to consider the possibility that they stand not only for letters but also for syllables. Likewise, I had to ask whether the ciphertext (or the plaintext behind it) is mostly made up of consonants, as in Turkish, Hebrew, or a stenography system.

In ciphers that code syllables, the symbols themselves are often complex—that is, syllable-like or two-digit numerals. In a modern syllable-coding matrix cipher, for example, the word "reinforcements" could be encoded in two different ways, again creating isologues (see table 1).

plaintext: r ei n fo r ce m en t s
ciphertext: 94 31 56 71 94 44 09 35 13 92
plaintext: re in f or ce m ent s
ciphertext: 98 28 74 59 44 09 39 92

	6	0	4	3	8	1	7	5	9	2
8	a	l	ad	al	an	and	as	at	b	2
4	c	3	ce	co	d	4	da	de	di	e
3	5	ea	ec	ed	ee	ei	el	en	ent	er
7	es	et	f	6	fi	fo	g	7	h	8
2	hi	ht	i	9	in	ing	io	ir	is	it
0	j	0	00	k	l	la	le	ll	m	ma
5	n	nd	ne	ng	ni	nt	o	on	or	ou
9	p	q	r	ra	re	ri	ro	rs	rt	s
1	se	si	st	t	ta	te	th	ti	tion	to
6	tw	ty	u	ur	v	ve	w	x	y	

TABLE 1 | A modern syllable-coding cipher. Reproduced from US Department of the Army, *Basic Cryptanalysis* (Field Manual 34-40-2), 5–7.

This system is simple and concise; its key does not extend beyond half a page—and yet an enormous amount of effort is required to break or even to recognize this type of code. Frequency analysis, vowel tests, and word-pattern analysis all fail here. The script of the Rohonc Codex suggests that a meaningful unit is sometimes one symbol, sometimes two, other times three or even four. In a complex syllable-coding method, this is exactly the case. Frequent repetitions may also be explained by assuming that single signs may stand for syllables and double signs for single letters. A word with the structure ABAB may simply be a double letter, if, for example, AB stands for *e*.

I tried to mark possible boundaries of character combinations with a pencil in a photocopy of the codex, looking for single and complex signs. What slowed this process down was the fact that for every unit I hoped to delimit, I soon found something that thwarted the pattern. For example, the two characters of some frequent double signs, such as the renowned symbol for Christ, can also be found independently. Afterward, I presumed that certain characters, such as the reverse *c* or the symbol resembling the Greek letter delta, may be the row coordinates in a matrix cipher like the one above, while the symbols standing after them could be the column coordinates. In such a system, if there are many columns but only a few rows, the row coordinates will be much more frequent than the column coordinates, effectively confusing an ordinary kind of frequency analysis.

Finally, I imagined a complex system that involves nomenclators (signs standing for specific names and places), letters, and syllables, indicated by two symbols each in a matrix system, and perhaps nullities, too. The key to such a system would easily fit on two A4-size pages, it would be user-friendly, and coding and decoding would be simple. The large number of symbols in the Rohonc Codex, the new signs appearing out of nowhere (the nomenclators of new people and new places in the story), the concise form of "words" or combinations of symbols that seem to make up one linguistic unit (coded syllable by syllable, not letter by letter), and the two different INRI signs (suggesting that a letter may have two equivalents in the cipher key)—all of these things suggest that the encryption of the Rohonc text may be based on just such a complex system. Despite all of my efforts, I failed to reconstruct such a system, though I could not exclude this possibility.

Two more possible scenarios remain. The first is that the text contains no vowels. Since this possibility is essentially the same as if it were a stenography—that is, shorthand—it will be discussed in the next chapter. The second option is that of a pure code system. In that case, nomenclators outnumber letter signs to such a degree that the system is basically made up of signs (or sign combinations) for whole words. Such a code system, which could only be deciphered with a great deal of luck and an enormous amount of work, poses problems similar to those of an artificial language, so these two things are discussed together below.

The Rohonc Codex resisted my efforts at cryptanalysis. One reason for this is that I couldn't limit the language of its plaintext to the required degree. Nor could I identify the code system, although the four most likely candidates seem to be the homophonic cipher, the syllable-matrix table, the code (whole word) system, and the consonant-only theory. Whenever I was close to giving up, I reminded myself that the condition of the codex strongly suggests that this book was extensively handled and read. If it is not a hoax, we may assume that the codex was used by members of a community who were able to read it continuously, probably after spending some time studying the encryption system on a one- or two-page key. If this assumption is correct, then sooner or later, someone will crack the code.

CHAPTER 7

SHORTHAND SYSTEMS

STENOGRAPHY: AT ONCE FAST AND SECRET

When someone writes a text in shorthand (in other words, stenography),[1] she obviously has different—if not contrary—goals from someone who enciphers a text. Despite the obvious differences, shorthand systems and ciphers do have a lot in common. Trying to read an early modern shorthand text without its character table would be quite similar to looking at a ciphertext without its key. Solving such a shorthand text can become a real challenge, even though its author did not intend to hide its meaning.[2]

The helpless frustration of their readers is not the only similarity between ciphers and shorthand systems. Some of the literature on the two genres also overlaps,[3] and often the same authors designed both ciphers and shorthand systems. A typical example is the father of modern stenography, John Willis (not to be confused with the mathematician John Wallis, or with John Wilkins, an artificial-language designer). Willis believed that one positive feature of his stenographic system was that it was inaccessible to anyone unfamiliar with it.[4] Having introduced what he calls "stenographie or compendious writing," he gives the five rules of "stenographie or secret writing." His advice is to mix up the letters before coding them with his characters, to systematically change the position of the vowels (e.g., to replace *e*'s with *a*'s, *i*'s with *e*'s, etc.), to introduce particular symbols for certain concepts (based on some similarity), to assign different meanings to the same character in different places, and to mark the end of a sentence using a special symbol, so that it remains hidden.

Stenography and cryptography may also overlap when the person designing the artificial language cannot quite decide whether he is creating a cipher or a shorthand system. This is the case in a number of private diaries that were written in secret characters in the ill-concealed hope that coming generations would eventually solve them—for example, the biography of the nineteenth-century Egyptologist Simeone Levi and the secret diary of the Hungarian writer Géza Gárdonyi, both mentioned earlier.[5] Both were broken after a great deal of work.

The history of stenography goes back a long way. Shorthand symbols, also known as Tironian notes (*notae Tironianae*), were widespread in medieval chancelleries until the twelfth century. The term derives from Marcus Tullius Tiro (103–4 BC), a slave of Cicero who was responsible for recording the rapid-fire speeches of his master. Over the centuries, the system became more and more complex, and by the twelfth century it had come to comprise several thousand signs, including symbols for both letters and concepts. By the later medieval period, however, cursive Latin writing had become widespread enough, and abbreviations and truncations were becoming sufficiently known, that Tironian notes were less necessary and useful. Shorthand symbols practically disappeared, to the extent that in the nineteenth century medieval texts written in shorthand had to be broken using the methods applied in solving a cipher or a dead language.

Following the disappearance of Tironian notes, there is a gap in the history of stenography for a couple of centuries. Basically, it had to be reinvented in the sixteenth century, and three early modern authors are considered the new founding fathers of shorthand: Timothy Bright (*Characterie; an Arte of Shorte, Swifte and Secrete Writing by Character*, 1588), John Willis (*Art of Stenography*, 1602), and Thomas Shelton (*Short-Writing*, 1626—in later editions titled *Tachygraphy*, Greek for "speedy writing").

As shorthand systems became more popular, their interrelationships became more complicated, and different schools flourished. Over the past four centuries, as many as 450 methods have appeared in print in English-speaking lands alone, although these often borrowed principles and even actual characters from one another.[6] Certain authors patented their own shorthand systems. Many realized that shorthand can to a certain degree be used as a cipher—Isaac Newton used Shelton's method several times during his student years to record private information in his notebook (for example, his list of confessions).[7] Samuel Pepys also relied on Shelton's method in his famous diary, in which he recorded the events of the English Civil War between 1660 and 1670. This diary was decoded between 1819 and 1822 by one John Smith, a researcher at the University of Cambridge, making both Pepys and Shelton famous immediately. The diary is regarded as a significant document in the history of English literature.

A common feature of early modern shorthand systems is that easy-to-draw, linelike characters were substituted not only for the letters of the alphabet but also for syllables and frequent words, like articles and conjunctions. Vowels were often left out, and their position could only be guessed indirectly from the consonants that followed them, the width of the line, and the surrounding punctuation. Shorthand systems usually worked with 150–200 signs altogether, thus putting the codebreaker into a situation similar to that of breaking a homophonic cipher. Pepys even introduced characters with no meaning, similar to nullities in a cipher. Although there is very little literature on breaking texts in shorthand to which the key is lost, it is clear that the usual tools of cryptanalysis can generally be applied.[8] The task is made much easier, of course, when the language of the plaintext is known.

SHORTHAND AND THE ROHONC CODEX

Is one of the many shorthand systems designed in the seventeenth century to be found on the pages of the Rohonc Codex? In the hope of answering this question I took myself to the Réserve, the special department of manuscripts and old prints of the Sainte-Geneviève Library in Paris. This proved to be the best place to do this research, for it is there that the eleven-hundred-item library of Louis-Prosper Guénin, the great late nineteenth-century historian of shorthand, can be found. It is hard to imagine a place with a higher concentration of texts written in either Tironian notes or modern shorthand.

Having browsed through the collection, I found encouraging leads. Christian liturgy, the apparent content of the Rohonc Codex, is not terribly foreign to shorthand. Stenography was used as often to record prayers and other religious texts as to take meeting minutes. When the Reverend James Humphreys left his Massachusetts home in 1776 to fight for American independence, he was already familiar with Janes Weston's 1727 handbook, *Stenography Completed*. In the breaks between battles he could record his religious experiences using Weston's method, just as Reverend Alexander Ewing used John Byrom's system to document his personal prayers around 1780, while Reverend James Hawkes chose the shorthand method of Henry Barmby around the same time. In each case, the motive of hiding personal messages was at least as important as increasing the speed of writing.[9]

It was not unusual for a designer of a shorthand system to illustrate the advantages of his invention with the Lord's Prayer or an excerpt from the Bible.[10] In a somewhat later period, the complete Gospels were printed in shorthand. I viewed the Gospels in Taylor's shorthand from 1843 at the Réserve in Sainte-Geneviève.[11] In the one-page key at the beginning of the thin book, each letter is assigned a symbol, but the same symbol may also stand for a preposition, a frequent object, or the ending of a word, depending on its position (before a word, after a word, or on its own). The key also contains a list of symbols for the most frequent words (*and, the, so, to, out, to be, shall be, together*), and for religious terms (*heaven, Christ, Gospel, the world, cross*).

Another biblical fragment, this one from 1886, encodes the text of the New Testament using Isaac Pitman's system.[12] The word "chapter" appears on the pages of the book in English, and the numbers are recognizable arabic numerals. The basic elements of Pitman's code are simple, but the signs that are built from the base units are rather complex and are frequently repeated owing to the repetitive nature of the text. Looking at the pages of the shorthand Gospel without the character key, we would be lost. The symbols are combined together in composite forms, and it is hard to distinguish them even with the help of the key.

There are several arguments, however, against the theory that the Rohonc Codex is written in shorthand. Shorthand systems are economical and they primarily encode single letters, even if they do contain many abbreviations and word signs. The characters of the Rohonc Codex—those that appear to be basic units—are not simple strokes. On the contrary, most of them take time and effort to draw. That makes the writing process *slow*, whereas the essential feature of shorthand is that it is *fast*. Furthermore, the primary goal of shorthand systems is not secrecy but speed and utility, which is why they often preserve elements of the plaintext, such as recognizable numerals. This happens with page numbers, for example, though many shorthand systems do use signs to designate numbers. Stenography aims to teach the reader, not exclude him. This cannot be said of the Rohonc Codex. There are no page numbers, chapter headings, or numbering systems that hint at the cultural background of the codex. Finally, shorthand systems normally go from left to right, which is how right-handed people write quickly without smearing the ink. Writing from right to left, as in the Rohonc Codex, is impractical, another strike against the theory that the codex is written in shorthand.

It is true that there were early shorthand systems that employed boustrophedon (from the Greek for turning like oxen when they plow land)—that is, the writing of alternate lines in opposite directions, from left to right and then right to left. Some stenographic systems used for private purposes, such as diaries or prayers, were indeed secretive, and some did read from right to left, however impractically.

For this reason it would be premature to rule out the possibility that the Rohonc Codex is written in an early form of shorthand, one that proved too impractical to attain the status of the famous systems of Willis and Bright. And perhaps the codex is after all somewhat secretive, like the system "perfected" by Pepys, which even contained nullities. The possibility that the author was trying to put into practice a language of his own design leads to consideration of the next step in our investigation: artificial-language schemes.

_____

CHAPTER 8

ARTIFICIAL LANGUAGES AND CODES

MAPPING THE WORLD

Universal, perfect, artificial, or philosophical language schemes informed a number of different, sometimes contradictory projects in the early modern period.[1]

Some authors were in search of the particular ancient language that, according to the Bible, Adam used to name the objects of the world. Other language designers attempted to map what they viewed as the true structure of the world using a language in which the relationships between words reflected the putatively real arrangement of objects in the world. Still others took a practical approach and aimed to create a common writing system that the various nations could use to communicate with one another, reading a unified system of symbols in their own languages.[2] There were those who wanted to ease tensions between people of different religions in the hope that a common, well-built, logical set of signs could decide basic religious questions, or at least help reach consensus on common religious foundations. Some wished to exchange Latin for a less outdated language that scholars could use to communicate, while others worked on unifying the grammar of the various natural languages. A few authors made their book-long utopias more exciting by inserting languages that they invented, while others made up a language as part of a big hoax. Magic and angelic alphabets were meant to communicate with the spirits, while some people made up languages out of sheer intellectual joy and aesthetic pleasure.

This kind of activity did not end with the early modern period, of course. Today there are whole internet communities organized around designed languages.[3] Detailed manuals and language-design kits are available to guide users through the steps of creating a new language, to help them come up with an exciting grammar, aesthetic writing, natural-sounding vowels, and harmonic words.[4] When a newborn language is especially fine-tuned and aesthetic, the internet community enthusiastically congratulates the designer.

This activity is not all that eccentric today, nor was it in the past. Many significant actors in the history of philosophy and science created artificial-language schemes or wrote about their importance and the need for such schemes. I have already mentioned Johannes Trithemius and John Wilkins, and we may also add Athanasius Kircher, René Descartes, Isaac Newton, Gottfried Wilhelm Leibniz, Marin Mersenne, George Dalgarno, Joseph de Maimieux, Francis Lodwick, Cave Beck, the Hungarian György Kalmár, and another Hungarian who is better known for his other pursuits, János Bolyai.

According to Umberto Eco, the search for a perfect language is one of the few stories that begins at The Beginning. The problem of "the ancient language," which Adam used to name everything on earth, has intrigued us ever since the Lord confused the languages at Babel.[5] Although the events at the foot of the famous tower mark the symbolic start of this process, the really important figures at the cradle of the language-creating tradition were the inner circle of the Hebrew Kabbalah and—probably independently—the author of the *Ars Magna*, Raimundus Lullus, a thirteenth-century Catalan philosopher.

At the center of Lullus's concept was his thinking machine, which consisted of three rotating concentrically arranged circles that combined the letters symbolizing various concepts. The concepts behind the letters reflected the structure of the world. According to Lullus, the whole setup of the world could be traced back to a finite number of elements. Rather than search for a lost ancient language, Lullus designed a practical philosophical tool that would automatically answer all questions. In his view, if all people could receive the same answers to the problems of the world, all kinds of religious and philosophical disputes would be resolved. Heretics would then convert to Christianity and peace would reign throughout the world.

The Hebrew Kabbalah was not meant to be a perfect language, yet its combinatorial methods offered procedures for approaching absolute, hidden, infinite, and ineffable truth. In the system of the Kabbalah, transposing and mixing the letters of the alphabet could lead to knowledge of the hidden God. There is an especially strong connection in the Kabbalah between the elements of the world and those of language.[6]

The quest for a perfect language was long influenced by the Cratylus debate, from one of Plato's famous dialogues, which addressed the relationship between language and reality. Is there a natural, original language whose elements are directly related to named objects, one in which the names are based on the inner nature of things? Or did language merely evolve according to convention, and were names assigned arbitrarily, based on tradition and mutual agreement? The former position, the belief in an original correspondence between things and their names, paved the way for the quest to identify the perfect ancient language. There was much at stake: if one could discover the language that Adam used to name the things of the world, one would have complete knowledge. Many authors regarded Hebrew, or some uncorrupted ancestor of Hebrew, as the *lingua Adamica*, but there were more exotic views that identified it as being Chinese, Swedish, Irish, or the Antwerp dialect of Flemish. Only Hungarians knew as early as the fifteenth century that Adam and Eve actually communicated in Hungarian as they fled Paradise.[7]

Early modern artificial-language schemes aimed to build on a finite and transparent quantity of basic concepts. Their creators were often motivated by a desire to achieve religious harmony. Raimundus Lullus, Guillaume Postel, Athanasius Kircher, Tommaso Campanella, Giordano Bruno, John Amos Comenius, and Gottfried Wilhelm Leibniz all regarded the perfect language as the path to peace, something that would help relieve all kinds of conflicts, erase all differences, and in general restore harmony on earth. This desire was sometimes channeled into politics. Scholars often endeavored to convince rulers of the advantages of universal monarchies or ideal republics that could be based on a universal language.

Early modern thinkers were interested in the problem of listing all the possible combinations of the existing elements of language and the world.

Encyclopedias, museums, theaters, libraries, and catalogs alphabetically listing the phenomena of the world and revealing its structure flourished in
the sixteenth and seventeenth centuries. According to the French polymath
Marin Mersenne, if we combine all nineteen consonants and ten vowels of
the French alphabet in every possible way, we will have said everything that
can be imagined and have produced altogether 11,327,512,775,262,815,560,
000,000,000 words. The little-known Swiss mathematician Pierre Guldin
calculated in 1622 that if we created all the words possible by combining the
twenty-three letters of the Latin alphabet, and listed these in thousand-page
books, we would need such a huge library to house them all that it would
take up all the land in the world and some of the seas, and would reach the
top of the tallest buildings in existence.[8] The Borgesian situation: all the
world a library!

There were basically two ways of achieving the perfect language: either
find the lost ancient language or rationally construct a new, in some way
perfect, "logic." After spending some time searching for the original ancient
language, many philosophers took the second route. No longer attempt-
ing to reveal the real configuration of things, they worked on organizing a
clearly structured library that would be worth placing an object in. Once the
search for the ancient language had died down, two new directions opened
up: some thinkers emphasized the philosophical classification of concepts
in language design, while others tried to produce a practical, user-friendly
language. The common feature of philosophical-logical language schemes,
including John Wilkins's *Essay Towards a Real Character, and a Philosophical Language* (1668) and Leibniz's *characteristica universalis*, was that they
attempted to map the relationships between elements of the conceptual field
using the tools of logic. Such schemes outlined a logical artificial language
that was abstracted from human languages, the units and words of which
could be any arbitrary string of characters, although it was preferable that
their appearance reflected their position within the system.

Wilkins, for example, used 270 large pages to enumerate the phenomena of the world in a classification system that he deemed transparent. He
assigned letters to the parts and features of the world and graphic signs to
these letters as a next step. His sign for flame is Debα, but replacing the α with
an *a* yields Deba, which means comet. That is, the objects and phenomena
of the world that resemble one another in their essence (flame/comet) are
similar in their word forms as well. Every new thought or discovery can be

integrated into this classification of knowledge, except for untrue ones, for which there is no place in the system. Such a language projected the rules of human thinking and facilitated recognition, thus also serving to ease or even replace the act of thinking. The strings of signs were used to complete automatic operations and acted like a modern-day computer language.

Instead of aiming at conceptual, logical perfection, practical language schemes, including international auxiliary languages such as Esperanto, Volapük, and Ido, aimed at universal usage, or as close to universal as possible. To achieve this allegedly practical goal, easy pronunciation that was reflected in the structure of words was a top priority. Such languages strove to be easy to read, write, and speak. The form of words did not in the least attempt to reflect the position of the particular object in human knowledge. Word forms were modeled on Indo-European structure and sought to achieve a fair balance between French, Latin, German, and Spanish words.

Of course, artificial-language designers differed not only in their goals but also in the appearance of their languages. The words in the languages developed by Dalgarno and Newton[9] were made up of a combination of three or four Latin letters. The languages designed by Athanasius Kircher and Cave Beck, by contrast, were made up exclusively of numbers. Charles de Brosses, Joseph de Maimieux, Christian Berger, György Kalmár, and many others constructed their scripts using special characters (a method called pasigraphy, after Maimieux's work by a similar title), while John Wilkins and Johannes Becher used special symbols that corresponded to letters and numbers, which ultimately referred to Latin words. Certain modern artificial languages—for example, Alfa-kinetix—even apply moving characters, graphic signs in constant motion. These can only appear on a computer screen, of course, and reading them can soon make you dizzy.[10]

Most artificial-language designers have realized that creating a good language is not enough; it also needs to catch on. The most frequent method of promotion was to take a popular religious text, most often the Lord's Prayer, and translate it into the newly devised script, just as the designers of shorthand systems did. John Wilkins, Charles de Brosses, Pierre de Bernonville, and even the forger George Psalmanazar demonstrated the merits of their languages by translating this prayer, while Francis Lodwick did the same with a Bible passage. The Rohonc Codex fits into this environment, though the religious texts used to illustrate artificial-language schemes were usually one page long, not 450 pages.

The Lords Prayer.

FIG. 20 | The Lord's Prayer in the language devised by John Wilkins in his *Essay Towards a Real Character, and a Philosophical Language* (London, 1668), 395.

| 1 | 2 | 3 | 4 | 5 | 6 | 7 | 8 | 9 | 10 | | 11 |

Our Parent who art in Heaven, Thy Name be Hallowed, Thy

| 12 | 13 | 14 | 15 | 16 | 17 | 18 | 19 | 20 | 21 | 22 | 23 | 24 | 25 | 26 |

Kingdome come, Thy Will be done, fo in Earth as in Heaven, Give

| 27 | 28 | 29 | 30 | 31 | 32 | 33 | 34 | 35 | 36 | 37 | 38 | 39 | 40 | 41 | 42 | 43 |

to us on this day our bread expedient and forgive us our trefpaffes as

| 44 | 45 | 46 | 47 | 48 | 49 | 50 | 51 | 52 | 53 | 54 | 55 | 56 | 57 | 58 |

we forgive them who trefpaf against us, and lead us not into

| 59 | 60 | 61 | 62 | 63 | 64 | 65 | 66 | 67 | 68 | 69 | 70 |

temptation, but deliver us from evil, for the Kingdome and the

| 71 | 72 | 73 | 74 | 75 | 76 | 77 | 78 | 79 | 80. |

Power and the Glory is thine, for ever and ever, Amen. So be it.

Eee2 · - 1. (~)

The casual observer may wonder why marketing would be necessary, given the presumably user-friendly, transparent structure of any well-constructed philosophical language. The designers' intention was indeed to create a language that reflected what they saw as the true relations between things, or at least to create a transparent system of reference that is easy to understand, with easily recognizable elements and structure. In other words, artificial-language schemes should be easily decodable systems. In György Kalmár's system, for example, the sign for *death* differs from that for *life* in one small curl of negative meaning. Francis Lodwick used similar geometrical forms for groups of words that derive from one common verb.

Other "perfect" languages, however, are not so transparent. One may easily become discouraged by looking at the Lord's Prayer as coded by John Wilkins (fig. 20). It is true that distinctive signs are consistently used for distinctive nouns or conjunctions, but only a skilled codebreaker would be able to decipher their meaning, since the language is not sufficiently logical to aid the reader.

The first sign is the first-person plural possessive pronoun, the second is built on the character of business relations; the stroke on its left refers to a blood relationship, and the one on its right to direct ancestry. These two signs together mean Our Father, which is read *hai coba*.[11]

The decoding of such scripts is further hindered by the fact that some of the authors, rather than create a code out of their own native tongues, aimed to devise a language that transcended national language groups and boundaries. Francis Bacon, John Webster, and John Wilkins argued that the "real characters," the building blocks of their constructed languages, should stand not for words and letters but for things and concepts, denoting universal ideas that appear in every language, albeit in different forms.[12] The author's intentions and the end product do not always match, however, and we rightfully expect every early modern constructed language system to bear the marks of its inventor's native tongue.

There is one more noteworthy enterprise to consider. Its goal, again, was to facilitate communication, although it falls within the area of constructed scripts devised to suit already existing languages. In North America, beginning in the seventeenth century, missionaries and others (including some indigenous people) created writing systems intended for use by Native American tribes, each of which spoke a distinct language. One of these was the syllable-based language of the Cree, created by the Welsh Methodist minister John Evans in the 1840s in the hope of converting the various tribes of the Hudson Bay area to Christianity and still used by some Inuit tribes today. Another such script is Cherokee, devised in the 1820s by a Mr. Sequoya, which consists of syllable signs, letters, and other symbols. Perhaps the oldest constructed writing system of this type is the one used to note down the language of the Mi'kmaq Nation on the eastern coast of Canada. This system was created in the 1600s by Father Chrétien Le Clercq, a French missionary, using hieroglyphic signs. Prayer books and other religious texts survive in all three of these languages.

ARTIFICIAL LANGUAGES IN HUNGARY

Because the Rohonc Codex was found in Hungary, and because its only known place of origin, the Batthyány library, was a Hungarian noble family's book collection, it might be promising to look for clues about the writing in the codex among local universal-language schemes. This is not an easy task, because this issue is not the subject of much research; there are few studies of artificial languages that are explicitly Hungarian in origin, in spite of a respectable number of relevant sources.[13] The language schemes of György Kalmár, István Kovácsházi, István Gáti, and János Bolyai are often mentioned in major international monographs, which makes Hungary overrepresented in the field of constructed languages.

We know of no Hungarian artificial-language schemes from the time period in which we presume the Rohonc Codex was created. There are some, but they are from later times. The earliest and most famous Hungarian language designer was György Kalmár (1726–1781).[14] Kalmár's life was even more exciting and colorful than his language design. From Debrecen, Hungary, he traveled to Germany, the Netherlands, England, and even Russia and Constantinople. With his neglected appearance, torn jacket, missing teeth, scary, disfigured face, and sliding wig, he shocked people wherever he went. One German writer who had read Kalmár's works and thought highly of him simply could not believe that the wretched-looking vagabond knocking on his door could really be the Hungarian scholar.[15]

Given his extensive travels, it is not surprising that Kalmár published his universal-language scheme in three languages: Latin, Italian, and finally—in its most complete but still not perfect form—German.[16] Kalmár believed that a true universal language must be not a spoken but a written character language.[17] Although Kalmár's published volumes, the longest of which is only 120 pages, contain only part of the promised four hundred ideograms, it is apparent that he wished to create a language that would be logically and metaphysically universal and, like other past and future language schemes, would also facilitate world peace. Kalmár occupies a place of honor in the European language-constructing tradition, not only because he could write in several languages but also because he was also extremely well read. He must have been familiar with most of the existing language schemes, and he liked to compare his own scheme, which he believed united the advantages of those of Wilkins and Leibniz, to its predecessors.

The chart in the German edition of Kalmár's book tells us a great deal about his system (fig. 21). It contains a number of texts that are relevant from a cultural and scholarly perspective and written in the character language. The three sentences under item I are Newton's three laws of motion. Item II, a two-line quotation in Kalmár's language, is from the introduction to

FIG. 21 | The character language of György Kalmár, from his *Grammaticalische Regeln zur philosophischen oder allgemeinen Sprache, das ist, der Sprache aller Voelker Zeiten und Lebensarten* (Vienna: Kurzböck, 1774), table VI.

Copernicus's magnum opus, which was ten lines long in the original Latin version. Other quotations in this chart are from Grotius, Locke, Bernoulli, Galileo, d'Alembert, and Euler. The final quotation is from Plato.

A project developed by a slightly younger Hungarian stenographer, István Gáti (1749–1843), was closely tied to that of Kalmár.[18] Gáti devised three parallel plans: a shorthand system, a pasigraphy or universal writing system, and a universal language. When creating the universal language and the pasigraphy used to write it down, he, like Kalmár, referred to Locke, Leibniz, Wilkins, and Berger. For Gáti, the universal language and shorthand system (which he called "quick writing") were closely connected. His objectives, he wrote, were to demonstrate "how to take down spoken language, how to correspond with people who do not speak our language, and how to have conversations with these people upon meeting them." He said of his characters, "although these were not designed for quick writing, but for the universal writing system, they would also work for quick writing, since single symbols stand for complete words, such as r—writing, ˆ—sky, ˇ—sea, t—time."[19]

Gáti's aim was to create a kind of philosophical language that would classify the entire visible world. His words were made up of two or three letters, as in Newton's system. Water, for example, is *zsa* if in the sky and *zsá* if on the ground. Thus, *zsab* means cloud, and *zsáb* means sea. *Pa* is fruit tree, *pab* is apple tree, *pan* is plum tree, and *pagy* is orange tree. *Pá* is fruitless tree, *páb* cedar, *pán* ivory, and *págy* cypress.

Here are Gáti's terms pertaining to illness:

lu outer	*lú* inner
lub wound	*lúb* rash
luc ulcer	*lúc* weakness
lud blister	*lúd* dizziness
luf dirt	*lúf* fever
lug French disease	*lúg* toothache
luh piles	*lúh* cold
luk measles	*lúk* rheum
lul scabies	*lúl* ache
lum ulcer	*lúm* arthritis

The system is not complete, which leaves room for later additions. In Gáti's classification of malformations, for example, *lugy* stands for gray (hair), *luly* for bald, *luny* for squinty-eyed, *luty* for snub-nosed, *luts* for tight-fisted and hunched, *luds* for crippled, *lunaj* for blind, and *lúsz* for stutterer; however, *lúgy*, *lúty* and *lúts* do not yet have meanings.

He applied his system to several religious texts, including the Lord's Prayer, the Apostles' Creed, and the Ten Commandments, a trade letter sent from Vienna to India, a military message sent from Bonaparte to Cairo, and a master's certificate sent from Philadelphia to London. His version of the *Pater noster* begins as follows:

pub tofmá ton se böp tyabmá fid tóg lyudnöp
pater noster qui es in caelis nomen tuum sanctificetur
Father our who are in heaven name yours be sanctified

János Bolyai (1802–1860), the founder of non-Euclidean geometry, did not reach this level of detail, although we gather from his manuscripts that the goal of creating a universal language, a logical and easy-to-comprehend system of symbols, was a lifelong interest.[20] István Kovácsházi's design was much more complete, though also more superficial.[21] In his thirty-one-page Hungarian-French bilingual document, Kovácsházi describes a practical pictographic scheme that everybody can grasp and illustrates its use with none other than the Lord's Prayer (fig. 22).

Kovácsházi thought the ambitious scope of Gáti's project killed him. Kovácsházi's ambition was more modest—he only wanted to show that a universal language was possible; he left the details to others. Still, his plan was not as naïve or ridiculous as it may seem at first sight. Otto Neurath, a significant philosopher in the Vienna Circle in the early twentieth century, worked on a similar pictographic system that everyone could understand; the pictograms in modern airports (representing takeoff, toilets, and luggage) are a product of this project.

It is undeniable today that western European and Hungarian artificial-language schemes were utopian in several respects. As for the international auxiliary languages, Esperanto won over Volapük, but the spread of the English language made Esperanto unnecessary. As for the philosophical languages, people realized that there is no universal or perfect classification of the world that does not bear the marks of the author's cultural background.

FIG. 22 | The pictographic design of István Kovácsházi, illustrated with the Lord's Prayer. From István Kovácsházi, *Pantographia (egyetemes írás)* (Budapest: Franklin Társulat, 1877), 16.

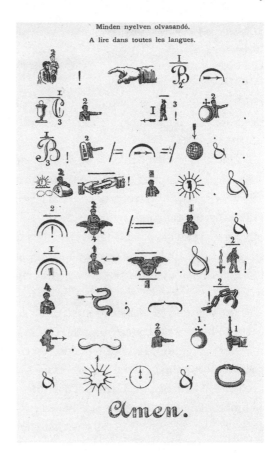

Viewed through the lens of a different tradition, such classifications, though meant to be perfect, may appear as comical as Jorge Luis Borges's famously ironic Chinese encyclopedia, where the beasts of the world are divided into the following categories:

1. Those that belong to the emperor
2. Embalmed ones
3. Those that are trained
4. Suckling pigs
5. Mermaids
6. Fabulous ones
7. Stray dogs
8. Those that are included in this classification

9. Those that tremble as if they were mad
10. Innumerable ones
11. Those drawn with a very fine camel hair brush
12. Others
13. Those that have just broken the flower vase
14. Those that, at a distance, resemble flies[22]

One must remember, though, that the ultimately abortive but in many ways fruitful philosophical program of the Vienna Circle was to some degree the heir to the tradition of artificial-language schemes, just as today's computer languages are that tradition's successful heirs. One lesson is clear: the more perfect an artificial language, the less universal it is. Limiting a language's field of application (as we do in programming languages) enables it to better answer questions emerging within its system. The less universal it is, the more perfect it becomes.

The tradition of artificial-language schemes has failed to bring world peace or to create the ideal communication that leads to inevitable truths that every religion and school of philosophy could accept. Not that we would consider these aims possible or necessary today. Still, this movement was not without benefits: we have gained a series of philosophical insights and a number of methods that have facilitated communication or improved calculation. They do not answer all questions, but they may be significant from a practical point of view.

INVENTED AND IMAGINARY LANGUAGES

Let us return now to international waters. A special class of artificial languages consists of languages that never existed and that do not even aid human communication. They do not aim to reconstruct the language of Adam, or to map the conceptual structure of the world, or even to facilitate communication between different language groups. So what other reason could there be for creating or recording such a language? There are several: (1) for talking to angels and demons, (2) to include in fictitious travelogues, (3) as part of a hoax, (4) for its own sake.

The magic character strings mentioned in chapter 5 belong to the first category. These often serve to name the planetary spirits and to aid communication with them. The "angelic alphabet" of the famous English scientist and magician John Dee (1527–1608), which Dee used to record messages from the angels he had conjured up, also falls into this tradition. The angels, by the way, dictated their messages via Dee's medium, the alchemist Edward Kelley, and they did so in reverse, as if they were reading them from a mirror. Another well-known Renaissance scholar-magician, Cornelius Agrippa von Nettesheim (1486–1535), set forth a list of celestial alphabets in his three-volume manual on the occult sciences.[23]

The constructed language of Saint Hildegard of Bingen, a twelfth-century mystic, can be considered the earliest artificial language in Europe. Her *Lingua Ignota*, or "unknown language," is actually a glossary of made-up nouns that begins with God (*Aigonz*) and the angels (*Aieganz*) and goes on to list various people, ailments, and body parts, which are followed by skin diseases, church offices, parts of nature, trees, herbs, flying animals, and finally insects, the last of which is the cricket (*Cauiz*). Hildegard used a secret alphabet to record her unknown language, which has a sign for every letter of the Latin alphabet. What the abbess was trying to achieve with this language is not known, but one may assume that it was born out of a godly inspiration and served mystical purposes.

Magical alphabets used to communicate with heavenly creatures have a long history in both the East and the West. They all have something in common, though: they assign one single character alphabet to the letters of the known alphabet, so basically they are all monoalphabetic.[24] They can easily be decrypted even if the reader has no access to the character table, and even if the language of the plaintext is not known.

Members of the second category are made-up languages that certain authors incorporated into book-length travel books to make them sound more exciting.[25] Thomas More, the author of the first utopia, who actually came up with the idea of "never-land" and who coined the term *utopia*, presented an alphabet to go along with the culture he had created, as did Gabriel de Foigny, Denis Veiras, and Simon Tyssot de Patot in the seventeenth century. These authors of lengthy fictitious travelogues and invented cultures devised either a language or a complete writing system.[26] The mid-twentieth-century linguist J. R. R. Tolkien followed in their footsteps in devising a detailed language system for the characters in his fantasy novel *The Lord of the Rings*.

The third category consists of the bluff languages that were used in some convincing hoaxes. The most famous example is that of George Psalmanazar (1679?–1763).[27] Psalmanazar appeared in Rotterdam in 1703 and claimed to have been born on an island called Formosa, today known as Taiwan.

Little was known in Europe at that time about this faraway land, its people, and their culture, so Psalmanazar quickly became something of a celebrity in London. He even managed to deceive the skeptical Royal Society in the end. Within a year he had published his book on his exotic homeland, where, according to his account, polygamy, cannibalism, and human sacrifices were common; where men learned classical Greek in school but walked around naked, with only a piece of gold or silver metal covering their genitals; where husbands had the right to eat their unfaithful wives; and where people traveled on horseback and camels and ate snakes. Psalmanazar constructed an alphabet and a complete language that he claimed was his native tongue, and he translated the Lord's Prayer and the Ten Commandments into it. His book was published in two editions in English and was also translated into French, Dutch, and German. For years, the author enjoyed the spotlight as a stranger from an exotic, faraway land, but news gradually arrived about the real Formosa that was not in the least consistent with Psalmanazar's account. Interestingly enough, some of this news came from the Jesuit missionaries who, Psalmanazar claimed, kidnapped him when he refused to become a Roman Catholic. When they contradicted the pretend native of Formosa, no one believed them. The "native," who was, by the way, blond, fair-skinned, and French by nationality, was eventually unveiled and had to admit, first to his friends and then to the public, that he had made the whole thing up.

Surprisingly, Psalmanazar did not disappear completely but became a relatively well known and respected linguist, with several publications on the Hebrew language and the Old Testament. Finally, he wrote a memoir in which he gave the full account of his story, but he arranged for it to be published only after his death—and even then he kept his real name secret.

Finally, there is a class of written languages that no one claims as belonging to an existing or even an invented faraway country. They exist simply because they were written down. The most famous examples are the Martian and Hindu languages of Hélène Smith, the Swiss medium who was Hungarian by origin, and the breathtakingly beautiful and mysterious Codex Seraphinianus created by the Italian architect Luigi Serafini. Hélène Smith (1861–1929) was a well-known spiritualist famous for conducting spiritual sessions. She believed herself to be a reincarnation of Marie Antoinette and claimed to communicate with Martians in a Martian language, and also that she was the heroine of Hindu stories in her dreams. She produced the Martian language in a trance state via so-called automatic writing. It consisted of twenty-one graphic signs, went left to right, and had no punctuation. Linguists studied the Martian writing carefully and identified a series of words of French, German, and Hungarian origin. *Writing*, for example, is *manir* in Martian, from the Hungarian word *iromány* (writing, document). In its sentence and word structure, Smith's language resembles Neo-Latin, and the scripts carry decipherable meaning, which the author translated into French herself.[28]

None of these statements can be made about the monumental codex by Luigi Serafini (1949–). This handbook contains curious imagery, scientific illustrations, and unreadable descriptions in an encyclopedic structure. It caused almost as much excitement in cryptographic circles as the Voynich manuscript. One thing about the Codex Seraphinianus distinguishes it from other unsolved writings: its author is known and alive.

Serafini created the book in the second half of the 1970s. The more than three hundred pages contain colorful "scientific" illustrations, but the written descriptions resemble no other known writing. The images usually depict absurd and surreal objects: trees that seem to be setting out for a walk, uprooting themselves, chemical reactions, human and animal parts, droplets of water that climb trees and fall back to earth. Its most famous series of drawings show how a copulating man and woman metamorphose into an alligator who—in the last image—climbs off the bed. Considered by many a parody of the Voynich manuscript, the codex contains eleven chapters, each with its own title page and table of contents. Based on the illustrations, the book seems to be structured thematically, with sections on biology, architecture, vehicles, flora and fauna, bizarre-looking two-legged creatures, traditions of historical faraway cultures, games—and one that perhaps gives a history of the writing of the codex itself.

The few copies of the Codex Seraphinianus were published at a high cost. Although it has been reprinted several times, book collectors are still willing to pay large sums for a first edition. Since its publication, it has intrigued and seduced many a codebreaker. Several statistical and linguistic studies have been made of it, and its writing system has been the subject of not a few master's theses. Some researchers, after much effort, have been able to identify the codex's numeral system, but the letters have withstood all analysis to date. After much conjecture about whether the text is a cipher, an artificial code language, or perhaps a hoax, the mystery was revealed by the author himself. At a meeting of the Oxford University Society of Bibliophiles in May 2009, Serafini admitted that the codex is indecipherable.

It is a so-called asemic writing, or a type of automatic writing in which the author produces a textlike flow of characters with no intention of sharing any specific intelligible content. Although the text is unintelligible, it produces an experience similar to reading. The text created by Hélène Smith was somewhat similar, but whereas Smith claimed that she learned this language by talking with Martians in a trance state, Serafini does not associate his writing with anything supernatural. Someone in the audience at the Oxford meeting recorded his words:

> The book creates a feeling of illiteracy which, in turn, encourages imagination, like children seeing a book: they cannot yet read it, but they realise that it must make sense (and that it does in fact make sense to grown-ups) and imagine what its meaning must be. . . . The writing of the Codex is a writing, not a language, although it conveys the impression of being one. It looks like it means something, but it does not; it is free from the cage of a language and a syntax. It involves a visual process, not a linguistic process.[29]

This, however, has not discouraged researchers of the codex. Recently, I received a dissertation via email. Its American author had examined the writing of the codex with visual-recognition software and argued that this method could work with the statistical analysis of other unsolved writings such as Rongorongo (used on Easter Island) or Indus Valley script. The author asked me if the Rohonc Codex should also be examined with this method.

An interesting question! Could the Rohonc Codex belong to this or any of the other categories described above? Is the Rohonc Codex an automatic writing? Did the author produce the codex in a trance state, creating a visual series of symbols devoid of meaning? Was it created with the intent of facilitating communication among different languages? Is it an invented hoax language to go along with a made-up culture, or perhaps a tool for communicating with angels? Is it simply the case that, as with Linear B, once it is finally solved, we will only smile at how wrong we were in all our misguided guesses? Or is it like the Voynich manuscript, which may never tell us whether it is a hoax or an intelligible text? The answer lies in a thorough study of the symbols.

CHAPTER 9

METHODS OF CODEBREAKING

AN ARTIFICIAL ROHONC LANGUAGE?

Does the Rohonc Codex contain a constructed language similar to those discussed in the previous chapter? We have seen that not even the most famous and most philosophical languages are easy to make sense of without a key: their graphic signs are not always easy to recognize, and even their structure does not always reflect the transparent logical system behind it. The languages of Kircher and Wilkins, however finely constructed, could only be solved by a lucky and very talented codebreaker. It is plausible that the Rohonc Codex was written in an artificial language that was not meant to be secret; all that happened was that the key—the table of symbols and the description of the system—got lost. Even the lack of character spaces on many pages is not a direct indication of intentional encryption. Although most artificial languages do not hide spaces (why would they?), neither do they go out of their way to emphasize them more than they are indicated in the Rohonc Codex.

If, however, the codex is an artificial-language system, what genre does it belong to? We can probably exclude the angelic language type of Hildegard of Bingen, John Dee, and Cornelius Agrippa, the Martian texts of Hélène

Smith, and the languages of imaginary travels, since these, as noted above, are all monoalphabetic and contain fewer than thirty different signs. It is possible, of course, that a text written with this kind of monoalphabetic alphabet may withstand codebreaking attempts, especially if the plaintext is written not in a natural language but in an artificial one known only to its author (and possibly to a small circle of his associates). However, one would still be able to carry out letter-frequency analysis and other cryptographic tests. Afterward, one could map the characteristics of the open language at least partially—the languages of Psalmanazar, Smith, and even Dee and Kelley reveal a lot about the grammatical structures of their designers' native languages. The problem is that this approach does not work in the case of the Rohonc Codex.

We had better examine the more comprehensive artificial-language schemes for an answer. Historians distinguish between purely philosophical (a priori) languages and those that are based on existing (a posteriori) languages. The Rohonc Codex is probably not a quest for the lost ancient language of Adam or even a detailed philosophical language like that of Wilkins, which offers a full classification of the world.[1] It is more likely to be one of the early seventeenth-century attempts to create a common language that would be more practical than perfect.[2] As we have seen, these focused on producing character writing that all literate people could read in their own languages, hence the name "universal character" or *escriture universelle* (universal writing). Such a writing system requires a fat dictionary, in which a large number of characters and character combinations denote the basic concepts and the verbs, as Francis Bacon and John Wilkins noted very early on.[3] The most famous of the many projects of this kind is Francis Lodwick's *Common Writing* (1647), but some of these designs were never printed. Scattered correspondence indicates that around 1630 three Frenchmen, Jean le Maire, D. P. Champagnolles, and someone named des Vallées; an Englishman, Philip Kinder; an Irishman, one Reverend Johnson; and a Swedish writer named Benedict Skytte, all working independently, came up with a version of an artificial language in which the words were represented by symbols. Sadly, none of these have survived—perhaps they were not all completed, or perhaps they are lost or still lying in oblivion in a manuscript collection somewhere.[4] Champagnolles's language scheme—which must have been sophisticated, because the author was able to transcribe the whole of Homer's *Iliad* and dedicate it to King Charles I—remained unpublished, because the inventor's widow wanted too much money for its publication. The language known as *Wit-spell* by the Reverend Johnson was not printed because no press was willing to undertake the preparation of its characters. The Rohonc Codex could be any one of these six projects! Or even a seventh one.

But chronology may undermine this possibility. As we have seen, the documented history of artificial languages in central Europe does not begin until much later than the supposed creation of the Rohonc Codex. In addition, these languages, with the exception of Kalmár's, are a lot more primitive than the system used in the codex (and even Kalmár's is incomplete). Nevertheless, there may have been similar designs in the region that ultimately remained unpublished. Or perhaps the Rohonc language was created in western Europe in the early seventeenth century, either by an author whose other books are known today or by someone of whom we have never heard.

Now, we would need to see the dictionary of this language—that is, the list enumerating the character combinations and matching them to words, concepts, frequent conjunctions, and pronouns. For each language scheme that has survived in its entirety, the dictionary contains several thousand items; Wilkins's, for example, is three hundred pages long, and each item refers not to an individual letter of an actual natural language but to a word in an artificial language. This means that we should not approach such a writing system as if it were a cipher, in which the letters of the plaintext are encrypted independently of the meaning of the words and code words (i.e., nomenclators) are few. Instead, we should consider it a code language, in which complete words and concepts predominate and letter-based encryption is rare or absent altogether. In short, we should apply the tools of codebreaking rather than those of cryptanalysis (frequency analysis, vowel identification, word-pattern analysis, and so on).

What, then, is codebreaking, and what are its tools? Codebreaking aims to transform the code language into a plaintext by partially or completely reconstructing the code key.[5] This key consists of a table or dictionary ranging from a single page to hundreds of pages in length. It tells us which word corresponds to which symbol, number, or letter combination. The longer the code key, the safer the code, but also, as we have seen, the less practical it is. Because of the nature of codes, they may often be broken only incompletely: although most of the text can be decrypted, certain symbols may remain unsolved. In a cipher, only twenty-five to thirty letters (or 150 syllables) need to be identified, and as we progress the work speeds up and the solution becomes complete. A code, however, may contain as many as two to three thousand items, some of which may remain unknown forever.

Groups of signs that fall at the beginning and end of the text are crucial in codebreaking. As noted earlier, they may contain a salutation, an address, a signature, or a similarly stereotypical text fragment that is easy to guess. Codes, like ciphers, may also contain signs for letters, which allow the encryption of words that are not in the code key—though words spelled letter by letter remain a minority (if they were the majority, we would call the text a cipher that also contains a few code words). Words that must be decoded letter by letter offer an excellent entry point into a code, as they often start and end with a specific sign, one that indicates the need to switch from code to cipher and back again. The "start spelling" (that is, letter by letter) and "end spelling" symbols are often the most frequent in the text, and they behave oddly: they occur intermittently. If these special signs can be identified, the text between them can be deciphered using the tools of cryptanalysis. Texts that employ this method often use isologues, groups of signs that stand for the same word—as we saw in chapter 6. Isologues are the different forms a word takes in a given text: once a word is decoded letter by letter, for example, then it is substituted as a whole by a code. Identifying isologues is a major breakthrough.

Apart from these relatively easy standard methods, however, the process of codebreaking is tedious. It basically involves writing down each code group and then examining the preceding and following code groups, looking for patterns. In other words, it is a statistical analysis applied on the code groups (the words), with special emphasis on their position and environment. In military codes, numbers frequently stand out, as they typically start with 0, 1, or 2. Punctuation marks, conjunctions like *and* and *but*, definite and indefinite articles, and similar words are also relatively frequent, so they are easy to spot for an experienced codebreaker, but not for a casual amateur. A professional codebreaker will count how many times a given code group occurs at the beginning, middle, and end of a message. If, for example, a given word occurs frequently between (any) two other code groups (and not, for example, at the end of a sentence), and if many different code groups surround it, then it is likely to be a conjunction (most often the word *and*).

Decoding words is obviously the most difficult part of the job, and it requires a skill different from deciphering. While the latter demands mathematical and statistical knowledge, codebreaking needs more of a sense of languages: the skill to recognize which words are likely to fall before and after a word whose meaning is perhaps already known. Guessing is frequently necessary: the codebreaker assigns a particular meaning to a given code and then tests this hypothesis. If the assigned meaning seems to be correct, how does that affect the meaning of other symbols? In the end, the pieces of the puzzle either fall into place, and the codebreaker moves on, or she must start again. The codebreaker will be able to guess correctly, however, only if she is thoroughly familiar with the rules of the given language. If this is a constructed language, she must explore its grammatical features. If she is lucky, she may even recognize the characteristics of the designer's native tongue, which he has almost certainly, if unintentionally, incorporated to some degree in his constructed language.

After this point, it is not easy to suggest algorithms that will always work: the codebreaker becomes increasingly skilled through practice, and her work resembles magic in many ways. And this magic cannot be performed in a day, and often not even in a week.

It was this magic that I tried to apply to the script of the Rohonc Codex. After spending weeks in libraries looking for historical analogues in old manuscripts, trying to determine what kind of artificial languages originated around the time the codex was written, I decided to avoid the library for an hour each day, find a quiet place, and pass the time studying the characters. In this way, whether I could make any sense of the codex or not, I hoped that something would jump out at me, something I could follow up on.

I made a list of the most frequent character combinations and ran a contextual study on them. I put each one on a separate sheet of paper, where I also recorded all the signs that precede and follow them. I kept an eye on signs that obviously start and end sentences, inscriptions that appear on the illustrations, and the signs before and after the "snakes" that appear to separate different sections of the text. As noted above, I looked especially closely at the signs that occur before, after, and between the signs identified as Christ and Pilate, which seem likely to be conjunctions, suffixes, or function verbs.

I often felt that I was on the right track. I identified a sign that may stand for *and* and another that is probably an article. Some symbols proved to be interchangeable, so they must have the same meaning. Others proved to be undividable composite signs. I was beginning to "learn" the words of the codex, which behaved differently in different positions. I became familiar with their characteristics even though I did not yet know their meanings.

Like other would-be codebreakers of this codex, I noticed that complete pages are sometimes repeated, with only slight modifications in the characters. Perhaps the author got bored and simply copied a previous section verbatim, or perhaps the same text was always coded with the same technique. This is one more reason to believe that we are not dealing with a homophonic text and should not bother looking for isologues.

I spent a long time comparing the strings of characters that occur before and after the illustrations, since they are often repeated with few differences.[6] I copied these similar strings, one under the other, on graph paper, in the hope that a common feature would stand out. Many details led me to believe that the groups of signs really do stand for whole words and not merely syllables, as I had previously assumed.

BREAKTHROUGH

While I was doing my best to "read" the codex, I learned that others were also working on it, and with considerably more success. Allow me now to share the story of two Hungarian researchers who worked first separately and then together on this project. The breakthrough came as a result of this cooperation.

In chapter 4 I introduced the codebreaking attempts of Gábor Tokai, an art historian and the head of the numismatics section of the Hungarian National Gallery. Although his field is the twentieth century, he is equally interested in earlier historical periods. Since his youth, he has been attracted to ancient unsolved languages, which explains why he began to work on the Rohonc Codex. His results were first published around the time I was finishing the Hungarian version of this book.[7] I was glad to see that he had discovered a series of characteristics that I had also recognized, but also a great many that I had not. Tokai did what real codebreakers must do: he bravely assigned meanings to certain groups of codes and then tested them. One of his most important results was that he assigned the names of the four evangelists to four regularly changing symbol groups he had identified. Although his initial hypothesis had to be altered (two evangelists "switched codes"), this was a major step in the codebreaking process.

It is interesting how Tokai's approach converged with, and at the same time differed from, that of the other codebreaker, Levente Zoltán Király. While Király identified symbols and groups of symbols with the help of a computer, Tokai often relied on gut feelings and his outstanding visual memory. He virtually memorized the 450 pages of the codex. Naturally, his intuitive insights alone would not have taken him far, and reasoning, confirmations, and disconfirmations always followed his intuitions. For example, he mapped the numeric expressions of the codex and began to construct the author's calendar system in the hope that this would allow him to guess the contents of the book. In addition, Tokai put in order the loose and mixed leaves, which had fallen out and were probably put back in the wrong order, before the pages were numbered. Since several dozen pages had almost certainly ended up in the wrong place this way, rearranging them was an important step in the process.

Király, a theologian and at the time a graduate student at the Reformed Seminary of Budapest, originally worked on identifying the structure and syntax of the codex's language. Through meticulous analysis, he found a number of features that others had missed. He tried to identify sentence-separating signs and studied the large number of short and long repetitions in general. By studying these repetitions, he also arrived at rearranging the scrambled leaves—parallel with but independently of Tokai. Király (again, like Tokai) also analyzed the numerals and identified the ordinal "first," which frequently appears next to several other numbers— for example, on the first of the three stone tablets in the illustration on folio 13v, where the ordinals "second" and "third" can be seen on the other two tablets. Király thus proved that the tablets are numbered. His most impressive achievement is an enormous chart arranging into columns a piece of text that appears forty-four times and follows more than half the illustrations, always with some slight variation (fig. 23).

This chart—which is similar to one that Tokai made with paper and pencil—is useful for many reasons. Semantic units can be visually distinguished, which is a great help, as word boundaries are not clearly visible in the Rohonc text. Prepositions can also be identified, and word order can be analyzed. And the chart reveals strong patterns. In the third section of text (counting from the right), one of four different symbol groups may appear in a given position. Since the signs following them seem to be numbers, the four symbol groups could be nicely substituted with the names of the

A comparative table of repeated symbol strings from the Rohonc Codex, with page references and index numbers in the right-hand columns:

	page	no.
	9v	4.
	14v	6.
	16r	8.
	21r	9.
	54v	22.
	59v	26.
	63v	27.
	65r	28.
	66v	29.
	68v	30.
	69r	31.
	71r	32.
	72r	33.
	79v	37.
	87v	41.
	88v	42.
	90v	43.
	92v	44.
	95r	45.
	97v	46.
	104v	51.
	107v*	52.
	109v	53.
	113r	54.
	128r	57.
	130r	58.
	134r*	60.
	158v	65.
	163r	66.
	167r	67.
	170r	68.
	175v	70.
	177v	71.
	182v	72.
	191v	74.
	193v	75.
	195v	76.
	197r	77.
	199r	79.
	202v	80.
	205v	81.
	210r	82.
	211v	83.
	213v	84.

FIG. 23 | A comparative study of the most frequent repetitions in the Rohonc Codex. From Levente Zoltán Király, "Struktúrák a Rohonci-kódex szövegében: Helyzetjelentés egy amatőr kutatásról," Theologiai Szemle 54 (2011–12): 89.

four evangelists; the numbers might then refer to specific biblical text. And indeed, the illustrations next to these symbols seem in most cases to match the biblical stories to which the strings of characters refer![8]

This was as far as Tokai and Király, working independently, had gotten by December 2010, when they began their collaboration. Together, they finalized the correct order of the loose leaves. They identified each of the references that contained the name of an evangelist and a biblical locus. Király created a password-protected codebreaking platform online into which he encoded the script of the whole codex. This enabled the pair to test any new hypothesis quickly, as the software automatically lists all the places where a given symbol combination appears, together with its environment.

When they got to the stage where they could read most of the codex continuously, Király and Tokai were generous enough to share their results with me. Having achieved no breakthrough myself, I had grown tired by now of trying to break the code. Although I was disappointed, of course, that I was not the one finally to break the code, I was also greatly relieved that the responsibility and duties that come with solving it had fallen on someone else.

So I studied Tokai and Király's solution.[9] Having poured so much time and energy into this project myself, it was a strange experience, and I would like to share it here.

A solution is proved correct when other people, in possession of the solution, are able to read the encrypted text. This happens naturally in the case of ciphers: having reconstructed a cipher key, the codebreaker shares the solution and lets others test it on the given ciphertext. If their solution is different from his, or if he does not provide a cipher key at all, he may be accused of not actually having deciphered the cipher.

Codes are a different matter. Confirmation or disconfirmation of a solution is far more complicated, and more circular. Suppose, for example, that I claim to have solved a text and state that it is composed of a finite number of code groups, and that these code groups refer to nouns, written one after the other, following no known grammatical rules. In that case, my theory can hardly be refuted. Using my key, any coded text can be decoded as a series of nouns. I may think that I have found the solution to the text, in spite of the fact that anyone could produce a different series of nouns that would also "map perfectly" onto the code text. So others could just as easily "prove" that the text is actually a list of demon names or alchemical terms. I would strongly suggest avoiding this approach to breaking unsolved languages, not because it does not work but because it works so easily. If we are clever about pairing words with symbols, we may easily get a prayer text that looks somewhat authentic, though perhaps unstructured and lacking any grammar. But well-chosen swear words could just as easily fit the symbols, making a meaningful albeit vulgar text.

Athanasius Kircher fell into this trap when he read the Egyptian hieroglyphs as if they were symbols of a code language. He gave Egyptologists a never-ending source of fun when he deciphered a combination of hieroglyphs, which today is thought to stand merely for "monarch," as "Osiris, creator of fertility and all vegetation, whose great potential is given to him from Saint Mophta from the heavens."[10]

The Rohonc Codex is a good example of this phenomenon. According to Tokai and Király's solution, which I read in manuscript form in 2011, the codex is written in a code language, some kind of artificial-language scheme. Therefore, we should look for words, or at least parts of words, not sounds or letters. Tokai and Király believe that the text is religious, containing stories from the Gospels, biblical paraphrase, and prayers. The kind of trap that Kircher walked into had been set. Did these two researchers manage to avoid it?

They were clearly aware of the danger. Let me quote their methodology—published years later—at some length:

> The principles of our criteria and method of codebreaking may seem banal to the reader, but we must emphasize them because of the bad reputation gained by the amateur researchers of the codex. Furthermore, as many examples in our next paper on the "wobbliness" of the code will show, the writing system is far from being simple and clean. We must affirm that these results are not due to methodically deficient research but to the writing itself, which was analyzed with painstaking care and strictness.
>
> We demand that one symbol signify one thing, and whenever there is any digression from this principle—either by more symbols signifying one thing or one symbol signifying more things—it must be sufficiently supported by argument. Our case is difficult because the codex has codes signifying words of a language, and words behave less regularly than letters. In every natural language the presence of homonyms and synonyms creates ambiguity. Yet we demand that even this amount of ambivalence in our proposed solution be supported by evidence.[11]

In order to decide whether Tokai and Király's solution is correct, we must first learn, to the extent possible, the vocabulary these codebreakers offer. Then, as a second step, we must accept the peculiar linguistic characteristics that they attribute to the language of the codex. We must provisionally trust their theory as we study and absorb this new, proposed language, in order to make a sound judgment of whether it works. All of this requires a great deal of time and effort, but eventually the first signs of confirmation begin to appear, just as they did for Tokai and Király.

The dictionary that Tokai and Király came up with consists primarily of words that have one meaning, and they carry that specific meaning every time they appear in the text. Disturbingly, however, the same symbol combinations sometimes stand for different things, as in the case of the words *us*, *man*, and *you*, which are represented by the same combination of symbols. Similarly, one symbol combination can mean either *I am* or *you are*; another stands for both *yours* and *ours*. Such peculiarities could, of course, be explained if we knew the language of the plaintext, but we still don't have that vital piece of information. In spite of such ambiguities, the authors are certain that they are on the right track: "The core of our reading has such strong inner and outer evidence that we may affirm that it stands beyond doubt."[12]

Once the lexicon has been grasped and the odd grammar accepted, more and more of the codex opens up, and clear-cut confirmations become visible. Using the vocabulary learned from one part of the codex, it becomes possible to read other parts. It becomes clear that the biblical references in the codex correspond to New Testament passages: the same numerals that appear in the Gospels can be identified in the codex: five barley loaves and two fish, twelve baskets, ten thousand talents, two hundred denarii, and so on. But numbers are not the only things that can be identified: Tokai and Király also find the Lord's Prayer in the codex, and also the *Ave sanctissima Maria* and *Ave Maria gratia plena*, along with literal quotations from such Gospel texts as the parable of good and evil fruit in Matthew 7:17–18.

Problematic sections remain, however, and certain grammatical peculiarities are left unexplained. Despite every encouraging detail, the natural language behind the text remains hidden and unknown. We know which group of codes stands for *table*, but we do not know what language the author of the codex spoke when uttering this word.

Other codebreakers have begun to respond to Király and Tokai's analysis and partial decoding of the Rohonc Codex. As noted above, it takes time for a new solution to gain acceptance. Király and Tokai's solution, as in the case of Linear B, will be either proved or refuted by the academic world, and this is a slow process. Eventually, other codebreakers will either confirm Király and Tokai's work or conclude that it is a self-proving theory and cannot be confirmed. Cryptology expert Nick Pelling has expressed his ambivalence about their solution on his website, and the reader comments on his analysis also represent differing attitudes.[13] Pelling is convinced about the codicological results but skeptical about Király and Tokai's interpretation of specific numbers, the calendar, and also about the code system. He points out that the decryption does not go below the level of individual words, and that the plaintext language is not identified.

Regardless of the mixed reviews, I am convinced that Király and Tokai have correctly decoded the Rhonc Codex, for two reasons. First, the authors shared with me their unpublished results about the grammar of the script very early, and I was able to check details that were not available to the public. Second, I had access to the entire dictionary, which could not be incorporated into their *Cryptologia* article because of its length. I trust that further parts of the solution, which should soon be in print, will meet with the approval of most codebreakers. Until then, anyone interested in confirming or disconfirming the solution, or simply working out another solution, can consult Tokai and Király's website, called "Der Rechnitzer Kodex," where the complete digitized codex can be "read." This "online tool," the authors explain, "is created to facilitate the research of this extraordinary manuscript." The glossary—that is, the dictionary, which of course is essential—is searchable. The user can type any Rohonc word into the search box and be provided not only with its meaning but also with its occurrences and various grammatical settings throughout the codex.[14]

The story does not end here. In the past decade or so, as Király and Tokai were working on their solution, others engaged in parallel efforts to crack the Rohonc Codex. In 2014, on the basis of his earlier investigations, Marius-Adrian Oancea created a detailed website on the codex. While performing a systematic analysis of its symbols, Oancea prepared a digital version of the codex and attempted to identify the names of the four evangelists and the saints. He suspects that the script (at least for the most part) is based not on words but on an alphabet and that this alphabet is inspired by Old Hungarian runes. His decryption is not finished, and—to my knowledge—no native Hungarian speaker can read the codex along these lines (Oancea himself knows only basic Hungarian). Still, after Tokai and Király's decryption (details of which were not available in English in 2014), Oancea has prepared the most thorough systematic analysis of the codex to date.[15]

Similar in size and depth is Delia Hügel's "book (in hypertext) about a book (in ciphertext)," which the author activated in early 2014. Hügel offers the most detailed analysis of the drawings in the codex, providing a large range of analogues from art history. She also shares her results on the statistics and meaning of the symbols and the variations of the sign combinations and identifies the names of Christ, Pilate, Moses, and so on in the script. While she offers no final answers, her web-book is definitely worth reading.[16]

If I were starting my research today, that's where I would begin: with the work of Gábor Tokai and Levente Király, Marius-Adrian Oancea, and Delia Hügel.

CODA

As a result of the research carried out in recent decades by various people around the world, we can conclude with absolutely certainty that the Rohonc Codex is not a faked document parading as an ancient Hungarian text. Given that it employs many characters in typical and recurrent combinations and omits obvious spaces between words, it does not look as if the author was trying to convince the reader that it was written in a natural language.

Neither does it seem to have been created for the purpose of duping a book or antique collector into paying a fortune for it. That is, it does not appear to be a hoax. With its simple gray binding and somewhat childish drawings, it looks neither valuable nor particularly mysterious (apart from the fact that it is written in unknown characters). Compared to the Voynich manuscript, with its decorous appearance and illustrations of mystical galaxies, exotic botanical creatures, and nude women, the Rohonc Codex is a rather drab affair.

The suggestion that the author was a lunatic, his work insane gibberish, does not hold up, either. To begin with, it was probably written and read by several people, which would require the cooperation of a group of nutty maniacs. And even if it was written by a single author, the recurrent semi-grammatical structures indicate that it is not the result of random or spontaneous automatic writing; the author clearly appears to have followed certain established rules in the process of writing. The consistent use of certain character combinations seems clearly to rule out both the hoax theory and the lunatic theory.

If we accept that the codex is an intentionally constructed text, we are faced with the question: what kind of text? Its strong repetitions (low level of entropy) make it an unlikely candidate for a polyalphabetic cipher. It is more probable that the same character is always assigned to the same letter, word, or concept. We can thus reasonably conclude that the Rohonc Codex is either a shorthand stenography or some other kind of consonant-based writing, a homophonic or syllable-based cipher that contains code words, or an artificial language.

All three hypotheses can be challenged. Unlike extant early modern shorthand systems, the codex contains characters that were not produced by quick and simple strokes but were carefully drawn—a slow and impractical method completely inconsistent with the principles of stenography. Unlike the many extant ciphers from the sixteenth and seventeenth centuries, the codex is not a diplomatic document but a religious text. It is also much longer than the historical analogues, which were short letters, not long books. And although a wealth of artificial languages have survived from the early modern era, and while, like the codex, they often do not signal word boundaries and were frequently applied to religious texts, none of them is as early or anywhere near as long as the Rohonc Codex.

As discussed in previous chapters, we might plausibly wonder whether the codex could be the product of some combination of these three possibilities. For example, could it be written in a form of shorthand that resembles an artificial language containing primarily code words, or a cipher that mostly encrypts whole words, along with some letters? These possibilities led us to apply the methods of codebreaking—but an essential prerequisite of codebreaking is knowledge of the plaintext language, and only further historical research on the possible origins of the codex can yield that information (if we are lucky).

We arrive at a paradox. Whatever coding system was used to create the Rohonc Codex, it must be outdated, yet it has thwarted our best efforts to crack it completely open. It is outdated, since none of the pre-1838 encrypting technologies have been able to resist the science of cryptology—theoretically. Knowledge of the context (the language, genre, and historical background of the plaintext) should provide enough information for today's codebreakers to break this code. Without that information, even the best professional codebreakers working around the clock for years on end may fail to find a complete solution—yet, theoretically, it is not an impossible task. And still, although a simple substitution code is considered unsophisticated, and although the system is breakable in theory, it may continue to withstand all of our attacks.

William Friedman, the outstanding American codebreaker who cracked a number of difficult war ciphers, was once asked if he considered the unsolved Beale ciphers, which allegedly contain the location of hidden treasure in the American West, a hoax. "On Mondays, Wednesday and Fridays," he replied, "I think they are real; on Tuesdays, Thursdays and Saturdays, I think they are a hoax."[1] János Jerney, who in 1842 became the first scientist seriously to attempt to decipher the Rohonc Codex, imbibed the same cocktail of enthusiasm and despair. "As for me," he wrote in 1844, "I reach for it eagerly, turning the pages again and again, making reliable copies of the original. I look at it several times, I make comparisons, imagine I have reached useful results. Then I get tired of reading it and set this rarity aside again."[2]

Nearly 180 years later, anyone who spends even a couple of hours with the codex will understand Jerney's emotional seesaw. There are just enough regularities in the text to keep the codebreaker eagerly engaged, and enough irregularities (the repetition of complete pages, the appearance of new signs and variations just when he thinks he has perceived a clear pattern, uncertain limits of the symbol combinations) to drive him mad.

After 180 years, however, the Rohonc Codex seems finally to have surrendered to the codebreakers. Gábor Tokai and Levente Király have made a compelling case that the codex is a breviary-like volume consisting primarily of New Testament readings. They and other researchers will surely refine these findings, until the curve of the broken code asymptotically approximates the imaginary line of the perfect solution, possibly never reaching it.

In closing, I would like to borrow the somewhat theatrical words of the man who first published the Beale ciphers in 1885. This is how he explains why he decided to write the story of the ciphers rather than solve them: "In consequence of the time lost in the above investigation, I have been reduced from comparative affluence to absolute penury, entailing suffering upon those it was my duty to protect, and this, too, in spite of their remonstrances. My eyes were at last opened to their condition, and I resolved to sever at once, and forever, all connection with the affair, and retrieve, if possible, my errors. To do this, as the best means of placing temptation beyond my reach, I determined to make public the whole matter, and shift from my shoulders my responsibility."[3] I could say much the same about why I decided to redirect my own energies toward writing the story of the Rohonc Codex and my attempts to crack its code, once I realized that I could not solve it. It is up to the reader to decide whether it was worth the trouble.

The Images in the Rohonc Codex

To number the pages (and therefore the illustrations), I relied on nineteenth-century folio numbering. According to the direction of the writing, we begin numbering at the back of the codex, so that the last page is numbered 1. Every left-hand page gets a number and an *r*, being a recto page, and the opposite page (the back of the previous sheet) receives the previous number and a *v*, meaning that it is the verso of the previous one.

The folio number is followed by a description of the depicted scene, whether biblical or not, and the biblical references are indicated, which help identify the given scene. The identification of the scene is based on iconographical features.[1] Further—and sometimes different—identifications are possible on the basis of the decryption of the written text, for which see Király and Tokai's article.

4v	Probably an evangelist and an angel.
5r	Unidentified scene: an angel, a person leaning forward, and behind him another person, holding a sword, perhaps Abraham preparing to sacrifice Isaac (Genesis 22).
8v	Unidentified scene: Christ and a person with a bishop's miter in front of a church.
9v	Unidentified scene: Christ speaking to two people.
13v	Two people standing between three inscribed stone tablets on a mountain, possibly Moses and the Tablets of the Law on Mount Sinai (Exodus 31:18), though it is not clear why there are three tablets rather than the biblical two. According to Levente Zoltán Király, the tablets are numbered; the characters on them read, from right to left, 1, 2, and 3.

14v   Christ entering Jerusalem on the left: here we see Christ on a donkey holding a cross and with a halo; people are spreading their cloaks on the ground before him, as in many depictions of this scene. Next to Christ is the palm tree that is being stripped so that the leaves may be laid in his path (Matthew 21:8–9; Mark 11:1–11; Luke 19:30–38 [no palm fronds here]; John 12:12–15). In the building on the right, which stands for both Jerusalem and the Temple, Christ drives the money changers out of the Temple with a whip in his hand (Matthew 21:12–13; Mark 11:15–17; Luke 19:45–46; John 2:14–16). Christ's entry into Jerusalem and the driving out of the money changers occur in this chronological sequence only in the synoptic Gospels, but only the Gospel of John mentions a whip made of cords.

15v   The sacrifice of a pigeon in the Temple (Luke 2:24).

16r   On the left: the Annunciation (Luke 1:26–38). An angel in the Temple passing an enormous lily to Mary. Words issuing from God can be seen above Mary, and the dove symbolizing the Holy Spirit lands on her. Unidentified scene on the right: an angel talking to someone, probably Joseph, who was visited by an angel in his dream after the Annunciation and told not to leave the pregnant Mary (as he had secretly planned to do) (Matthew 1:20). The angel and the person stand on a mountain, a clear contradiction of the iconographical tradition. This is odd, and it undermines the identification of this illustration.

21r   The Epiphany, with the Magi adoring the newborn Christ, the Star of Bethlehem above them (Matthew 2:1–12).

24v   Unidentified scene: possibly the Transfiguration, found exclusively in the synoptic Gospels (Matthew 17:1–13; Mark 9:1–12, Luke 9:28–36), though it is more likely to be a baptism scene (cf. 116v).

26r   The Crucifixion, angels collecting Christ's blood in two chalices. The first and last characters above Christ's head (normally INRI) differ.

28v   Jesus arrested in the Garden of Gethsemane, with an olive tree. On the left, Christ and the disciples kneel in prayer; on the right, soldiers come from Jerusalem (Matthew 26:36–56; Mark 14:32–50; Luke 22:48–53; John 18:1–11). At bottom left, possibly Peter among the servants at the house of the high priest (Matthew 26:69–75; Mark 14:66–72; Luke 22:54–62; John 18:15–27). There is a character on the right, above the city and below the sun, that seems to refer to Jerusalem in other pictures (cf. 65r, 109v).

29r   Probably Judas and Christ (with a cross on the halo) at the Last Supper, both reaching for the bread at the same time (John 13:26) (cf. 92v, 205v).

34v   Decorative border, with Christ and an unidentified animal. Perhaps the Temptation of Christ (Matthew 4:1–11; Mark 1:12–13; Luke 4:1–13).

37v   Unidentified scene: perhaps a martyr who is thrown into water, or perhaps the biblical scene in which the Roman soldiers come for Christ and the disciples are still asleep (cf. the description of picture 28v). The sun and another star are featured, and a combination of characters above the soldiers (and perhaps Christ) appears in other pictures too (cf. 44r).

40v   Christ standing before Pilate, a soldier behind Christ. Pilate is probably holding a scepter with lilies (Matthew 27:11–26; Mark 15:1–15; Luke 23:1–25; John 18:28–40). The scene is framed with an ogee arch, a finial, and turret motifs. The double characters above the two figures probably refer to Christ and Pilate.

42r   Unidentified scene: probably Christ and Pilate, Pilate perhaps washing his hands (Matthew 27:24).

44r   The flagellation of Christ (Matthew 27:26; Mark 15:15; John 19:1). The character combination above the soldiers is the same as in picture 37v.

44v   A snake or eel before the last line of text (cf. 144v, 192v, 205v).

45r   Christ wearing the crown of thorns, before Pilate and with a soldier behind him (Matthew 27:27–31; Mark 15:16–20; John 19:2–5).

49r   Christ carrying the cross (Matthew 27:31–33; Mark 15:20–22; Luke 23:26; John 19:16–17).

51r   Crucifixion (Matthew 27:35–36; Mark 15:24–28; Luke 23:33; John 19:17–19). The INRI sign is not very clear here.

52r   Symbolic representation under the cross: two altars, each with a priest. It may refer to the contrast between the classical Old Testament (synagogue) and the New Testament (ecclesia), or to Eastern and Western Christianity.

54v   The events following the death of Christ, top: the *Noli me tangere* scene where Mary Magdalene meets the risen Christ, whom she mistakes for the gardener. After she recognizes Jesus, she wants to touch him (John 20:14–17). To the left, the empty tomb with an angel, who is perhaps rolling the stone away from the mouth of the tomb. To the right, perhaps John, who gets to the tomb first and to whom the angel shows the burial clothes (Matthew 28:1–8; Mark 16:1–8; Luke 24:1–12; John 20:1–18).

56v   Christ appears to the apostles (Mark 16:9–14; Luke 24:36–43; John 20:19–23).

57v   The Harrowing of Hell.

59r   A typical depiction of the Resurrection with Christ ascending, surrounded by the soldiers who watched over the tomb, flag in Christ's left hand.

59v   Probably Christ and the two disciples on the road to Emmaus. When Christ breaks the bread, the disciples recognize him, after which he disappears (Mark 16:12–13; Luke 24:13–35).

63v   Scene with Doubting Thomas. The key on the left symbolizes the fact that Christ appeared to Thomas through a locked door (John 20:26–29).

65r   Christ looking down on Jerusalem, with the usual symbol that indicates Jerusalem above the city (similar to 28v, 109v).

66v   Biblical parable, either about the good and bad fruit (Matthew 12:33), or about the cursing of the fig tree (Matthew 21:18–22; Mark 11:12–14, 20–26).

68v   Another parable, or Christ talking to a disciple.

69r   Biblical parable, probably the Parable of the Sower (Matthew 13:3–23; Mark 4:3–20; Luke 8:4–15).

71r   See 69r.

72r   Christ and the disciples at the Last Supper (Matthew 26:26–29; Mark 14:22–25; Luke 22:19–20).

75v   Probably the Ascension, with Christ's legs hanging from the cloud (Mark 16:19; Luke 24:50–51).

78v   The Tree of the Knowledge of Good and Evil at the bottom with the four rivers above, an angel on either side, and above the rivers three people, the middle one Christ, perhaps standing on a chalice. The figures on either side of Christ may be the other two figures of the Holy Trinity, the Holy Spirit on the left and God the Father on the right. The character combination above Christ is similar to the ones on 44r and 37v.

79r   Apocalyptic scene.

79v   Unidentified scene: at top (in heaven?), Christ and a similar figure approach a cross, below them a sun emitting strange rays, while beneath the sun four kneeling or standing figures turn toward a fifth one.

83r   The structure of the world: heavenly spheres appear to be rotating the earth with a cogwheel; two angels are moving the heavens. Below and around the earth (with hell beneath it) are heavenly spheres and planets, also cogs, with heaven at the top.

83v   Unidentified scene with the same cogwheels as 83r.

84r   Jesus and Peter walking on water (Matthew 14:22–33; Mark 6:45–52; John 6:16–21).

87v   A person (or God) is dictating, while another figure (an evangelist?) is taking notes.

88v   Christ talking to two people.

90v   Unidentified scene: Christ and a king?

92v   Unidentified scene: two people (one of them Christ?) around a dinner table, perhaps Christ and Judas dipping bread into the dish at the Last Supper (cf. 29r, 205v).

95r   Unidentified scene: Christ and someone else holding a scepter-like object, standing in a valley or above a river.

97v   Unidentified scene: Christ and someone else standing between two cities.

98v   A depiction of Calvary?

100r   Unidentified scene: the rays of the sun falling on a kneeling person.

101v   Unidentified scene: Christ talking to a kneeling person, with the sun above them.

103v   Unidentified scene: two people (one of them Christ?) are holding something (a lily?) together.

104v   Unidentified scene: Christ and a crowned person talking to each other.

107v   Unidentified scene: Christ giving something to a person, next to a city.

109v	Christ looking down on Jerusalem with a sword before him, two of the usual characters that probably stand for Jerusalem above the city (cf. 28v, 65r).
113r	Unidentified scene: a crowned person kneeling as if praying to the sun (which has a human face), the rays of which almost reach him (cf. 134r, 167r).
116v	Baptism scene, very similar to the one on 24v: a child is held above a river on a hill, while several others stand around in a circle.
127r	Unidentified scene: angel at the head of a bed or tomb.
128r	Unidentified scene: two people standing in two boats on the water.
130r	Unidentified scene: two people hold something together, standing on two sides of a table, the sun above them.
132v	An angel, with an open scroll in hand, is descending to a crowned sitting figure. Perhaps the angel with the scroll of Apocalypse (Revelation 10).
134r	As on 113r, a crowned person kneeling, as if praying to the sun, the rays of which almost reach him. Instead of a face, however, the sun is accompanied by a character combination often seen in the codex that resembles the Hebrew tetragrammaton (cf. 167r).
137r	Decorative border with repeating lines, circles, and other strokes. Perhaps the branch of an olive tree?
141v	Unidentified scene: two people holding a round object, standing on a globe.
144v	A snake or eel at the end of the text (cf. 44v, 192v, 205v).
148r	Crucifix with a deer or similar animal in front of it.
152v	Three crosses with Christ and the two criminals.
154v	Unidentified scene: two people talking.
158v	Unidentified scene: three figures (one of them possibly Christ), above them a decorative sawtooth line.
163r	Unidentified scene: Christ, standing in front of the city, talking to someone.
167r	Angel in front of the sun (which has a human face), the rays of which almost reach him, with a man lying on the ground before him (cf. 113r, 134r).

170r   Christ standing before the Temple.

172v   Unidentified scene: two figures.

175v   Unidentified scene: Christ sitting beneath an ogee arch. In the corner above the picture there are two people (evangelists?).

177v   Unidentified scene: three people (Christ at center in the foreground) standing face to face with three other people, a city below them.

182v   A man, perhaps Christ, sitting at a table with a cross before him.

186v   Calvary beneath the sun and moon.

191v   Unidentified scene: someone lifting up a small child with two people around them, one on each side.

192v   A snake or eel closing the text at the bottom of the page (cf. 44v, 144v, 205v).

193v   Perhaps the Wedding at Cana (John 2:1–11), two people (Jesus and Mary?) sitting by the table with two others standing behind them.

195v   Unidentified scene: a teacher (Christ?) and three other people (children?).

197r   Cross with two people.

198v   Unidentified scene: an angel standing before a person.

199r   Unidentified scene: flower bending over a hill?

202v   Unidentified scene: two people.

205v   Unidentified scene: two people at a table (cf. 29r, 92v).

205v   A snake or eel closing the text at the bottom of the page (cf. 44v, 144v, 192v).

209v   Decorative drawing at the bottom of the page.

210r   Unidentified scene: Christ talking to two people.

211v   Unidentified scene: two people in a boat approaching a third one onshore, with the city behind them, a crescent on the church or temple. Perhaps Christ calling his first disciples, Simon Peter and Andrew, after the miraculous catch of fish (Luke 5:1–10).

213v   Unidentified scene: Christ talking to two people, with houses or windows behind them.

INTRODUCTION

1. We all have our favorite unsolved writings, of course. One list of nine items, including the Rohonc Codex, can be found at http://www.omniglot.com/writing/un deciphered.htm.

2. Reeds, "Friedman's Transcription of the Voynich Manuscript."

3. To single out just two books and two websites from the rich literature, see D'Imperio, *Voynich Manuscript*; Kennedy and Churchill, *Voynich Manuscript*; René Zandbergen's website http://www.voynich .nu; and http://www.ciphermysteries.com, a site by Nick Pelling.

4. The manuscript is kept in the Manuscript Collection of the Hungarian Academy of Sciences (MTA) under the call number K 114. A description of the codex can be found in Csapodi, *"Magyar Codexek,"* 109. The pages of the manuscript can be downloaded from the repository of the library at http://real-ms.mtak.hu/80.

5. The book was Kennedy and Churchill's *Voynich Manuscript*.

6. Láng, "Why Don't We Decipher?"

CHAPTER 1

1. Kelecsényi, "Egy magyar régiségkereskedő"; Kelecsényi, "Literáti Nemes Sámuel útinaplója."

2. Kelecsényi, *Múltunk neves könyvgyűjtői*, 248–75.

3. Várkonyi, *Thaly Kálmán és történetírása*; Várkonyi, "Elveszett idő";

Várkonyi, "Történészvita Zrínyiről 1868-ban."

4. Gaskill, *Poems of Ossian*; Gaskill, *Reception of Ossian*.

5. Kelecsényi, "Egy magyar régiségkereskedő," 320.

6. Szabó, "Régi hún-székely írásról."

7. National Széchényi Library, MS Fol. Hung. 1365/1–2.

8. Láng, "Invented Middle Ages."

9. Marton, *I. András király korabeli imák*.

10. National Széchényi Library, MS Fond 16/32. Kelecsényi had already found this note; see his "Egy magyar régiségkereskedő," 313.

11. See Literáti's diary in five volumes, National Széchényi Library, MS Fol. Hung. 3006, and his letters to Miklós Jankovich, National Széchényi Library, MS Fond 16/491–92.

12. Szabó, *Régi hún-székely írásról*, 123–24. Other, later scholars shared Szabó's assessment of the codex; see Fejérpataky, *Irodalmunk az Árpádok korában*, 3; Tóth, *Magyar ritkaságok*, 18–20; Jenő, *Magyar irodalomtörténete*, 1:43, 724–25; Csapodi, *"Magyar Codexek,"* 109.

13. As early as July 1842, János Jerney gave a lengthy lecture on the volume that was donated to the Manuscript Collection of the Hungarian Academy of Sciences in 1838 and officially catalogued in April 1839. This lecture was published two years later in Jerney, "Némi világosítások."

14. Winkel, "Poe Challenge Cipher Finally Broken"; Winkel, "Poe Challenge Cipher Solutions."

15. Kruh, "Basic Probe of the Beale Cipher"; Kruh, "Beale Cipher as a Bamboozlement."

16. Gillogly, "Beale Cipher."

17. King, "Reconstruction of the Key."

18. The Voynich Manuscript, Beinecke Rare Book and Manuscripts Library, Yale University, fol. 78r, par. 1, lines 1–5, quoted by Rugg, "Elegant Hoax?"

19. Landini, "Evidence of Linguistic Structure"; Rugg, "Elegant Hoax?"

20. Rugg, "Elegant Hoax?"; Schinner, "Voynich Manuscript."

21. Szabó, "Régi hún-székely írásról," 135–36.

22. Kelecsényi, Múltunk neves könyvgyűjtői, 248–75.

23. Hungarian Academy of Sciences, Manuscript Collection, Történl. 4-r, no. 38, pp. 132–35.

CHAPTER 2

1. Jerney, "Némi világosítások," 35.

2. Jülg's 1885 letter to Hunfalvy can be found next to the codex in the Manuscript Collection of the Hungarian Academy of Sciences.

3. Müller's letter is also at the Hungarian Academy of Sciences, Manuscript Collection, Ral K 801: 1/1885.

4. Szinnyei, Magyar irók, 9:954–55.

5. Némäti, Rohonczi Codex Tantétel; Némäti, Rohonczi Codex Ábéczéje.

6. Records of the Committee on Hungarian Linguistics, Hungarian Academy of Sciences, Manuscript Collection, Ral K 1568.

7. Nyíri, "Megszólal 150 év után a Rohonci-kódex?" For Ottó Gyürk's reaction, see Gyürk, "Megszólal a Rohonci-kódex?"

8. Enăchiuc, Rohonczi Codex. "Rohonczi Codex" is the Hungarian title of the book, in which Rohonczi is an adjective that follows nineteenth-century Hungarian spelling. Why the archaeologist and linguist gives the codex a Hungarian title when she considers it Romanian is a mystery.

9. According to Marianne Rozsondai, a former archivist at the Manuscript Collection of the Hungarian Academy of Sciences.

10. Enăchiuc, Rohonczi Codex, 22.

11. Ibid., 674. As noted, Enăchiuc gives these lines in French translation. The English translation is my own, as are all translations in this book unless otherwise noted.

12. Ibid., 8.

13. For Enăchiuc's French translation, see ibid., 669.

14. Ungureanu, "Nu trageti in ambulanta."

15. Előd Esztergály, editor's note, Turán 7–8 (2004–5); on how the solution came about, see pp. 6–7.

16. Singh, "Rohonci Kódex" (with Hungarian translation by László Bárdi); see also Singh, "Rövid ismertető a brahmi."

17. Varga, "Rohonczi Kódex."

18. Gyürk, "Hogyan fejtettem meg Gárdonyi titkosírását?"; Gárdonyi, Titkosnapló.

19. Gyürk, "Megfejthető-e a Rohonci-kódex?"

20. Locsmándi, "Rohonci Kódex," 47, 57.

CHAPTER 3

1. Briquet, Filigranes, 507–8; a long list of very similar watermarks can be seen on pp. 477–532. Briquet was a noted Swiss filigranologist who pioneered the use of dating paper by identifying its watermarks.

2. Among others, see Nickell, Pen, Ink, and Evidence; Nickell, Detecting Forgery; Nickell, Real or Fake?

3. Koltai, Batthyány Ádám és könyvtára, 256.

4. Béla, Körmendi Batthyány-levéltár; Koltai, Batthyány Ádám és könyvtára.

5. I am indebted to János Tamás Németh for this idea.

6. For a twentieth-century author trying to decipher a 1796 letter in shorthand as if it were a secret writing, see Gillogly, "Breaking an Eighteenth Century Shorthand System."

7. Singh, *Code Book*, 57.

8. Viterbo, "Ciphered Autobiography."

CHAPTER 4

1. The complete Rohonc Codex, including illustrations, can be downloaded from http://real-ms.mtak.hu/80.

2. I would like to thank art historians Dóra Sallay and Gergely Buzás for these observations.

3. Fol. 111 is not included in the illustrations, but it can be viewed at http://real-ms.mtak.hu/80.

4. Tokai published several articles on the codex and its illustrations and symbols in 2010–11. See Tokai, "Rohonci-kódex"; Tokai, "Első lépések."

5. Király, "Struktúrák a Rohonci-kódex szövegében."

6. For a reliable description of the Apocryphal Gospels, and a more provocative interpretation, see Pagels, *Gnostic Gospels*. See also Doresse, *Livres secrets*.

7. Nagy, *Apokrif evangéliumok*.

8. Turdeanu, *Apocryphes slaves et roumains*. Contrary to what its title suggests, this book also covers New Testament apocrypha.

9. Mead, *Pistis Sophia*.

10. Turdeanu, *Apocryphes slaves et roumains*; Bozóky, *Livre secret des cathares*; Ivanov, *Bogomilski knigi i legendi*; Ivanov, *Livres et legendes bogomiles*; Turdeanu, "Apocryphes bogomiles."

11. I am indebted to Ottó Gecser for calling my attention to the *Vita Christi* texts.

12. Johannes de Caulibus, *Meditaciones vite Christi*.

13. Ludolf von Sachsen, *Vita Christi*.

14. Wehli and Zentai, *Biblia pauperum*.

15. I am grateful to Veronika Novák for suggesting the possibility of the Book of Hours.

16. See, for example, Tenschert and Nettekoven, *Horae B.M.V. 158*, 2:237, 3:957.

17. Tokai, "Rohonci-kódex"; Tokai, "Első lépések"; Király, "Struktúrák a Rohonci-kódex szövegében"; and the article co-written by Király and Tokai, "Cracking the Code."

18. Dán, *Erdélyi szombatosok*, 11–26.

19. Koltai, *Atyák mondásai*, 188–89; Dán, "Martin Seidel's Origo."

20. For more about the Christian mission under Turkish rule, see Molnár, *Katolikus missziók*, esp. 364–71; Molnár, *Lehetetlen küldetés?*

21. Bennassar and Bennassar, *Chrétiens d'Allah*, 288; Burchill, *Heidelberg Antitrinitarians*.

22. See, among others, http://www.textweek.com/art/crucifixion.htm; http://jesusmarie.free.fr/index.

CHAPTER 5

1. This chapter is the abbreviated version of my overview of the early history of cryptography in my *Real Life Cryptology*, 31–49.

2. Bernhard Bischoff arranges the medieval methods in neat categories in "Übersicht über die nichtdiplomatischen Geheimschriften." For the history of secret writing in general, see Singh, *Code Book*; Kahn, *Codebreakers*; Wrixon, *Codes*, and the journal *Cryptologia*, which has created a place for publications on the history of secret writings since 1977.

3. See https://simonsingh.net/crypt ography/cipher-challenge/cipher-challenge -essays.

4. See http://www.characterfrequency
analyzer.com; https://www.cryptool.org/de
/cryptool2.

5. So far, six volumes have been
published in this series, edited by Mohamad
Mrayati, Yahya Meer Alam, and M. Hassan
at-Tayyan: *Al-Kindi's Treatise on Cryptanal-
ysis*; *Ibn Adlan's Treatise*; *Ibn ad-Durayhim's
Treatise*; *Ibn Dunaynir's Book*; *Three
Treatises on Cryptanalysis of Poetry*; and
Two Treatises on Cryptanalysis.

6. Kahn, *Codebreakers*, 106–88.

7. Alberti, *Treatise on Ciphers*.

8. Trithemius, *Polygraphiae libri sex*;
Trithemius, *Steganographia*.

9. Della Porta, *De furtivis literarum
notis*; Della Porta, *De occultis literarum
notis*.

10. Gustavus Selenus was the pen name
of the famous Duke August of Braunsch-
weig. Gustavus is the anagram of his real
name, Augustus, while Selenus is a
reference to his town, Lüneburg. Duke
August was the name-giving founder of the
Herzog August Bibliothek of Wolfenbüttel,
whose collection reflects its founder's
enthusiasm for cryptography. See Selenus's
exhaustive review on cryptology: *Cryptomenytices*. For more on Duke August, see
Strasser, "Noblest Cryptologist"; Strasser,
"Die kryptographische Sammlung Herzog
Augusts."

11. Vigenère, *Traicte des Chiffres*.

12. Falconer, *Rules for Explaining and
Decyphering*.

13. Selenus, *Cryptomenytices*, book 5,
chap. 19. A similar method can be seen in
Vigenère, *Traicte des Chiffres*, 238.

14. Selenus, *Cryptomenytices*, book 3.

15. David Kahn discusses the difference
between the theoretical methods described
in the literature and the actual practice of
enciphering, noting that there is "an air of
unreality" about the former. See his classic
monograph *Codebreakers*, 156.

16. Meister, *Die Geheimschrift*; Meister,
Die Anfänge; Pasini, "Delle scritture in

cifra"; Platania, "Polonia nelle carte del
cardinale"; Cecchetti, "Scritture occulte."

17. Devos, *Chiffres de Philippe II*;
Biaudet, "Chiffre diplomatique"; Speziali,
"Aspects de la cryptographie."

18. Devos and Seligman, *Art de
deschiffrer*.

19. Von Rockinger: "Über eine bayerische Sammlung"; Stix, "Die Geheimschriftenschlüssel."

20. Láng, *Real Life Cryptology*.

21. Kahn, "Man in the Iron Mask."

22. Bobory, *Sword and Crucible*, 10.

23. For a recent overview of the
sixteenth-century history of Hungary, see
Pálffy, *Kingdom of Hungary*.

24. Láng, *Real Life Cryptology*, 131–78.

25. Tusor, "Pázmány bíboros olasz
rejtjelkulcsa."

26. See Pálffy, *Kingdom of Hungary*.

27. Láng, *Real Life Cryptology*, 51–83.

28. Láng, "People's Secrets."

29. Hargittay, *Régi Magyar Levelestár*,
2:123, 146, 253, 258, 363. For more on this
subject, see Láng, *Real Life Cryptology*.

30. Kerekes, "Kémek Konstantinápolyban"; Kerekes, "Hírszerzés a XVI–XVII
században"; Kerekes, *Diplomaták és kémek
Konstantinápolyban*; Petercsák and Berecz,
Információáramlás.

31. Láng, *Unlocked Books*, chaps. 3 and 9.

32. Dresden, Sächsische Landesbiblio-
thek, MS N. 100, fols. 198r–200v;
Bibliotheca Apostolica Vaticana, MS Pal.
Lat. 1439, fols. 346r–47v.

33. Thorndike, *History of Magic*,
4:150–82; Birkenmajer, "Zur Lebensgeschichte."

34. Biagoli, *Galileo's Instruments of
Credit*.

35. Principe, "Robert Boyle's Alchemical
Secrecy."

36. Herzfelder, "Kolozsvári Czementes
János könyve."

37. Sándor, *Balassi Bálint összes művei*.

38. Haller, "Napló."

39. Zsigmond, *Naplói*.

40. Vadai, "Két XVII. századi titkosírás megfejtése."

41. Várkonyi, "Tájékoztatás hatalma."

CHAPTER 6

1. Cryptanalysis is a modern term, coined by William Friedman, the great codebreaker of the first half of the twentieth century.

2. Devos and Seligman, *Art de deschiffrer*, viii.

3. Pesic, "François Viète."

4. Devos and Seligman, *Art de deschiffrer*; see also Seligman, "Traité de déchiffrement."

5. Quoted in Friedman, *Military Cryptanalysis*, 1:1.

6. Ibid., 1:5.

7. Chadwick, *Decipherment of Linear B*, 67.

8. See, among others, Friedman, *Military Cryptanalysis*; Kahn, "Enemy Codes"; Safford, "Functions and Duties"; US Army, *Basic Cryptanalysis*.

9. Poe, "A Few Words on Secret Writing," 33.

10. These steps are listed in this order in several cryptology handbooks. See, for example, Friedman, *Military Cryptanalysis*, 1:7; US Army, *Basic Cryptanalysis*, 2–4.

11. Friedman, *Military Cryptanalysis*, 1:7.

12. "The would-be solver must possess a thorough knowledge of the language employed, not only from the point of view of vocabulary but also from that of a knowledge of all the peculiarities of its grammar, syntax and idiom." Kahn, "Enemy Codes," 168. The pamphlet was published in January 1918 by the Intelligence (E), Ciphers, and General Headquarters of the British Expeditionary Force, and it was intended to teach novice cryptanalysts how to crack German field codes during World War I. Kahn discovered the pamphlet in the UK's Public Record Office in London and

reprinted it in *Cryptologia*. See also https://doi.org/10.1080/0161-119591883863.

13. De Leeuw and van der Meer, "Turning Grille."

14. Friedman, *Military Cryptanalysis*, 1:102–4, 141.

15. US Army, *Basic Cryptanalysis*, 4–32.

16. For Sukhotin's algorithm, see Guy, "Vowel Identification"; Foster, "Comparison of Vowel Identification Methods."

17. Sassoon, "Application of Sukhotin's Algorithm."

18. Viterbo, "Ciphered Autobiography."

19. For more details, see Bauer, *Decrypted Secrets*.

CHAPTER 7

1. Stenography (i.e., shorthand) is not to be confused with steganography, which is the hiding of the message—for example, simply hiding the script by writing it by invisible ink.

2. See, for example, Gillogly, "Breaking an Eighteenth Century Shorthand System."

3. See, for example, Costamagna, *Tachiografia notarile*; Falconer, *Rules for Explaining and Decyphering*.

4. See Willis, *Art of Stenographie*, second (unnumbered) page. This is an exceptionally rare book. I used the copy of the Bibliothèque nationale de France, Réserve, V 30834.

5. See Viterbo, "Ciphered Autobiography"; Gárdonyi, *Titkosnapló*; Gyürk, "Hogyan fejtettem meg."

6. Sams, "Cryptanalysis and Historical Research," esp. 92.

7. Westfall, "Short-Writing."

8. Sams, "Cryptanalysis and Historical Research."

9. Ibid., 94.

10. Mavor, *Universal Stenography*, plates 3 (Lord's Prayer) and 5 (excerpt from the book of Job).

11. *New Testament . . . in Taylor's system.* This rare text can be found at Bibliothèque Sainte Geneviève, Paris, Réserve, delta 68.223.

12. *New Testament . . . in an Easy Reporting Style.* This rare text can be found at the Bibliothèque Sainte-Geneviève, Paris, Réserve, delta 68.226.

CHAPTER 8

1. Couturat and Leau, *Histoire de la langue universelle*; Couturat and Leau, *Nouvelles langues internationales*; Borst, *Der Turmbau von Babel*; Rossi, *Clavis universalis*; Knowlson, *Universal Language Schemes*; Pellerey, *Lingue perfette*; Slaughter, *Universal Languages*; Strasser, *Lingua Universalis*; Eco, *Ricerca della lingua perfetta*, published in English as *Search for the Perfect Language*.

2. See, for example, Beck, *Universal Character*; Kircher, *Polygraphia nova et universalis*.

3. For a useful summary on contemporary constructed languages, see Higley, *Hildegard of Bingen's Unknown Language*. One can also search online for conlang, Zompist Bulletin Board, and Conlanger Bulletin Board.

4. Pablo David Flores, "How to Create a Language?," http://www.angelfire.com/scifi2/nyh/how__all.html; Rosenfelder, "Language Construction Kit," http://www.zompist.com/kit.html.

5. Eco, *Search for the Perfect Language*, 7–10.

6. For a good overview, see Bloom, *Kabbalah and Criticism*.

7. Szilády, *Sermones Dominicales*, 1:xii–xiii.

8. Eco, *Search for the Perfect Language*, 140.

9. Elliott, "Isaac Newton as Phonetician"; Elliott, "Isaac Newton's 'Of an Universal Language.'"

10. See http://alfakinetix.blogspot.com and http://www.omniglot.com/writing/rotor.htm.

11. Eco, *Search for the Perfect Language*, 238–59.

12. Knowlson, *Universal Language Schemes*, 16 and 25. For Wilkins, see his *Mercury*, 109.

13. László Körmöczi and István Szerdahelyi did great studies in the last century that examined previously neglected sources, but their work is out of date. See Körmöczi, *Világnyelv kérdése*; Szerdahelyi, *Bábeltől a világnyelvig*.

14. János Bolyai is better known as a mathematician than as a designer of constructed languages. His scheme was not as detailed as Kalmár's and remains unpublished.

15. Hegedűs, *Kalmár György*, esp. 27–32.

16. Kalmár, *Praecepta grammatica*; Kalmár, *Precetti di grammatica*; Kalmár, *Grammaticalische Regeln*.

17. On Kalmár's philosophical language, see Hegedűs, "Kalmár György világnyelv tervezetének fogalomanyaga"; Balázs, "Kalmár György, a hazai nyelvbölcselet úttörője"; Balázs, "Kalmár György egyetemes írásnyelv tervezetének nyelvfilozófiai jelentősége."

18. Gáti, *Stenographiának I-ső könyve*. See also Gáti's unpublished works, "Stenographia, Tachygraphia, Pasigraphia" and "Steganographia vagy öszve vont írás."

19. Gáti, "Stenographia vagy öszve volt írás," fol. 1.

20. See Bolyai, "Üdvtan."

21. Kovácsházi, *Pantographia*.

22. Borges, "John Wilkins' Analytical Language," 231–32.

23. Agrippa von Nettesheim, *De occulta Philosophia libri tres*, 274.

24. Le Pape, *Écritures magiques*.

25. Knowlson, "Ideal Languages."

26. See, for example, Foigny, *Terre australe connu*.

27. On Psalmanazar, see Keevak, *Pretended Asian.*

28. Flournoy, *From India to the Planet Mars.*

29. Stanley, "To Read Images Not Words," 9, quoting Enrico Prodi.

CHAPTER 9

1. On the Wilkins lexicon, see Slaughter, *Universal Languages,* 112–24.

2. Knowlson, *Universal Language Schemes,* chap. 2.

3. Ibid., 53–56.

4. Slaughter, *Universal Languages,* 109–20; Lewis, *Language, Mind and Nature,* 24–42.

5. For more on codebreaking, see Kahn, "Enemy Codes"; Safford, "Functions and Duties"; US Army, *Basic Cryptanalysis,* chaps. 14–15; Friedman, *Military Cryptanalysis.*

6. For example, on fols. 59v, 63v, 65r, 68v, 69r, 71r, 72r, 88v, 90v, 92v, 97v.

7. Tokai, "Első lépések."

8. Király, "Struktúrák a Rohonci-kódex szövegében," 89–90.

9. Király and Tokai reveal their results in their co-authored article "Cracking the Code."

10. Quoted in Doblhofer, *Signs and Wonders.*

11. Király and Tokai, "Cracking the Code," 293.

12. Ibid., 295.

13. See http://ciphermysteries.com/2018 /06/03/kiraly-and-tokais-rohonc-codex -decryption. For a similarly mixed review, see the comments of cryptoblogger Klaus Schmeh at http://scienceblogs.de /klausis-krypto-kolumne/2018/06/03 /rohonc-codex-top-ten-crypto-mystery -solved-part-1-of-2.

14. See http://www.rechnitzer-kodex.hu.

15. See http://rohoncbyoancea.blogspot .com.

16. See https://rohonczcodex.wordpress .com.

CODA

1. Quoted in Clark, *Man Who Broke Purple,* 126.

2. Jerney, "Némi világosítások," 26.

3. Quoted in Singh, *Code Book,* 121.

APPENDIX

1. I am deeply indebted to art historians Gergely Buzás and Dóra Sallay for their help in identifying these drawings.

BIBLIOGRAPHY

MANUSCRIPT SOURCES

Bolyai, János. "Üdvtan" [Salvation theory].
Library of the Hungarian Academy
of Sciences, Microfilm 1120.

Gáti, István. "Steganographia vagy öszve
vont írás." National Széchényi
Library, MS Oct. Hung. 4.

———. "Stenographia, Tachygraphia,
Pasigraphia, Universalis Lingua
Rövid foglalatja." National Széchényi
Library, MS Oct. Hung. 467.

Literáti Nemes, Sámuel. Diary. 5 vols. National
Széchényi Library, MS Fol. Hung. 3006.

———. Letters to Miklós Jankovich.
National Széchényi Library, MS
Fond 16/491 (dated letters) and Fond
16/492 (undated letters).

Némäti, Kálmán. Rohonczi Codex Ábéczéje
[The alphabet of the Rohonc Codex].
Library of the Hungarian Academy
of Sciences, Manuscript Collection,
MS 884.

Records of the Committee of Linguistics.
Library of the Hungarian Academy
of Sciences, Manuscript Collection,
MS Ral K 1568.

Rohonc Codex. Library of the Hungarian
Academy of Sciences, Manuscript
Collection, MS K 114. The pages of
the manuscript can be downloaded
from the repository of the library at
http://real-ms.mtak.hu/80.

PRINTED SOURCES AND SECONDARY
LITERATURE

Agrippa von Nettesheim, Henricus
Cornelius. De occulta Philosophia
libri tres. Cologne: Soter, 1533.

Alberti, Leon Battista. A Treatise on Ciphers.
Torino: Galimberti, 1997.

Balázs, János. "Kalmár György, a hazai
nyelvbölcselet úttörője" [György
Kalmár, the pioneer linguist of
Hungary]. In Magyar Deákság:
Anyanyelvünk és az európai modell
[Hungarian literacy: Our native
tongue and the European model],
23–40. Budapest: Magvető, 1980.

———. "Kalmár György egyetemes
írásnyelv tervezetének nyelvfilozófiai
jelentősége" [The linguistic-philo-
sophical importance of the universal
language scheme of György Kalmár].
In Hermész nyomában: A magyar
nyelvtörténet alapkérdései [In the
steps of Hermes: Basic questions
of Hungarian language history],
450–67. Budapest: Magvető, 1987.

Bauer, Friedrich L. Decrypted Secrets:
Methods and Maxims of Cryptology.
Berlin: Springer, 1997.

Beck, Cave. The Universal Character by
which all the Nations in the World
may Understand one another's concep-
tions. London: Thomas Maxey, 1657.

Béla, Iványi, ed. A Körmendi Batthyány-
levéltár reformációra vonatkozó
oklevelei I, 1527–1625 [The charters of
the Batthyány archives in Körmend
I, 1527–1625]. Szeged: József Attila
Tudományegyetem, 1990.

Bennassar, Bartolomé, and Lucile Bennassar.
Les Chrétiens d'Allah: L'histoire
extraordinaire des renégats XVIe–
XVIIe siècles. Paris: Perrin, 1991.

Biagoli, Mario. Galileo's Instruments of
Credit: Telescopes, Images, Secrecy.
Chicago: University of Chicago
Press, 2006.

Biaudet, Henry. *Un chiffre diplomatique du XVIe siècle: Étude sur le cod. Nunz. Polonia 27. A. des archives secretes du Sant-Siège*. Helsinki: Annales Academiae Scientiarum Fennice, 1910.

Birkenmajer, Alexander. "Zur Lebensgeschichte und wissenschaftlichen Tätigkeit von Giovanni Fontana (1395?–1455?)." *Isis* 17 (1932): 34–54.

Bischoff, Bernhard. "Übersicht über die nichtdiplomatischen Geheimschriften des Mittelalters." *Mitteilungen des Instituts fur Österreichische Geschichtsforschung* 62 (1954): 1–27.

Bloom, Harold. *Kabbalah and Criticism*. New York: Seabury Press, 1975.

Bobory, Dóra. *The Sword and the Crucible: Count Boldizsár Batthyány and Natural Philosophy in Sixteenth-Century Hungary*. Newcastle upon Tyne: Cambridge Scholars Press, 2009.

Borges, Jorge Luis. "John Wilkins' Analytical Language." In *Selected Nonfictions*, edited by Eliot Weinberger, 229–32. New York: Penguin Books, 1999.

Borst, Arno. *Der Turmbau von Babel: Geschichte der Meinungen über den Ursprung und Vielfalt der Sprachen und Völker*. 6 vols. Stuttgart: Hiersemann, 1957–63.

Bozóky, Edina, ed. *Le livre secret des cathares: Interrogatio Iohannis; Apocryphe d'origine bogomile*. Paris: Beauchesne, 1980.

Briquet, Charles-Moïse. *Les filigranes: Dictionnaire historique des marques du papier des leurs apparitions vers 1282 jusqu'en 1600*. Paris: Picard, 1907.

Burchill, Christopher J., ed. *The Heidelberg Antitrinitarians: Johann Sylvan, Adam Neuser, Matthias Vehe, Jacob Suter, Johann Hasler*. Bibliotheca Dissidentium 11. Baden-Baden: Valentin Koerner, 1989.

Cecchetti, Bartolommeo. "Le scritture occulte nella diplomazia veneziana." *Atti del Regio Istituto Veneto* 14 (1868–69): 1186–211.

Chadwick, John. *The Decipherment of Linear B*. Cambridge: Cambridge University Press, 1958.

Clark, Ronald. *The Man Who Broke Purple*. Boston: Little, Brown, 1977.

Costamagna, Giorgio. *Tachiografia notarile e scritture segrete medioevali in Italia*. Rome: ANAI, 1968.

Couturat, Louis, and Leopold Leau. *Histoire de la langue universelle*. Paris: Hachette, 1903.

———. *Les nouvelles langues internationales*. Paris: Hachette, 1907.

Csapodi, Csaba. *A "Magyar Codexek" elnevezésű gyűjtemény* [The collection entitled "Hungarian Codices"]. Budapest: MTAK, 1973.

Dán, Róbert. *Az erdélyi szombatosok és Péchi Simon* [The Sabbath-keeping Christians in Transylvania and Simon Péchi]. Budapest: Akadémiai Kiadó, 1987.

———. "Martin Seidel's Origo et fundamenta religionis Christianae and Simon Péchi." In *Socinianism and Its Role in the Culture of the XVIth to XVIIIth Centuries*, edited by Lech Szczucki, 53–57. Warszaw-Lódz: Polish Scientific, 1983.

Della Porta, Giambattista. *De furtivis literarum notis vulgo de ziferis liber quinque*. Naples: Johannes Baptista, 1602.

———. *De occultis literarum notis, seu artis animi sensa occulte aliis significandi*. Strasbourg: Zetzner, 1606.

Devos, J. P. *Les chiffres de Philippe II (1555–1598) et du Despacho Universal durant le XVIIe siècle*. Brussels: Académie Royale de Belgique, 1950.

Devos, J. P., and H. Seligman, eds. *L'art de deschiffrer: Traité de déchiffrement du*

XVIIe siècle de la Secrétairerie d'État et de Guerre Espagnole. Louvain: Université de Louvain, 1967.

D'Imperio, Mary E. *The Voynich Manuscript—An Elegant Enigma.* Laguna Hills, CA: Aegean Park Press, 1978.

Doblhofer, Ernst. *Signs and Wonders.* Moscow: Eastern Literature, 1963.

Doresse, Jean. *Les livres secrets des gnostiques d'Égypte.* Paris: Rocher, 1992.

Eco, Umberto. *La ricerca della lingua perfetta nella cultura europea.* Bari: Laterza, 1993.

———. *The Search for the Perfect Language in the European Culture.* Oxford: Blackwell, 1995.

Elliott, Ralph W. V. "Isaac Newton as Phonetician." *Modern Language Review* 49 (1954): 5–12.

———. "Isaac Newton's 'Of an Universal Language.'" *Modern Language Review* 52 (1957): 1–18.

Enăchiuc, Viorica. *Rohonczi Codex: Descifrare, transcriere si traducere* [The Rohonc Codex: Deciphering, transcribing, and translating]. Bucarest: Editura Alcor, 2002.

Falconer, J. *Rules for Explaining and Decyphering All Manner of Secret Writing.* London: Printed for Dan. Brown, 1692.

Fejérpataky, László. *Irodalmunk az Árpádok korában* [Our literature in the ages of the House of Árpád]. Budapest: n.p., 1878.

Flournoy, Théodore. *From India to the Planet Mars: A Study of Somnambulism with Glossolalia.* New York: Harper, 1901.

Foigny, Gabriel de. *La terre australe connu (1676).* Edited by Pierre Ronzeaud. Paris: Société des textes français modernes, 1990.

Foster, Caxton C. "A Comparison of Vowel Identification Methods." *Cryptologia* 16 (1992): 282–86.

Friedman, William. *Military Cryptanalysis.* Vol. 1. Laguna Hills, CA: Aegean Park Press, 1980.

Gárdonyi, Géza. *Titkosnapló* [Secret diary]. Edited by Szalai Z. Sándor. Budapest: Szépirodalmi, 1974.

Gaskill, Howard. *The Poems of Ossian and Related Works.* Edinburgh: Edinburgh University Press, 1996.

———. *The Reception of Ossian in Europe.* London: Athlone, 2002.

Gáti, István. *A stenographiának I-ső könyve: A tachygraphia vagy szapora irás módja* [The first book of stenography: Tachygraphy or quick writing]. Pest: Trattner, 1820.

Gillogly, James J. "The Beale Cipher: A Dissenting Opinion." *Cryptologia* 4 (1980): 116–19.

———. "Breaking an Eighteenth Century Shorthand System." *Cryptologia* 11 (1987): 93–98.

Guy, Jacques B. M. "Vowel Identification: An Old (but Good) Algorithm." *Cryptologia* 15 (1991): 258–61.

Gyürk, Ottó. "Hogyan fejtettem meg Gárdonyi titkosírását?" [How did I solve the secret writing of Gárdonyi?]. *Élet és Tudomány* 24 (1969): 2211–16.

———. "Megfejthető-e a Rohonci-kódex?" [Is it possible to decipher the Rohonc Codex?]. *Élet és Tudomány* 25 (1970): 1923–24.

———. "Megszólal a Rohonci-kódex?" [Will the Rohonc Codex finally speak?]. *Theologiai Szemle* 39 (1996): 380–81.

Haller, Gábor. "Napló, 1630–1644." [Diary, 1630–1644]. In *Erdélyi Történelmi Adatok* [Transylvanian historical data], vol. 4, edited by Szabó Károly, 1–103. Kolozsvár, 1862.

Hargittay, Emil, ed. *Régi Magyar Levelestár* [Old Hungarian correspondence]. Budapest: Magvető Könyvkiadó, 1981.

Hegedűs, Béla. *Kalmár György (1726–?) világáról* [About the world of

György Kalmár (1726–?)]. Budapest: Argumentum, 2008.

———. "Kalmár György világnyelv tervezetének fogalomanyaga" [The concept in the universal language scheme of György Kalmár]. *Irodalomismeret* 6, nos. 1–2 (1995): 100–108.

Herzfelder, Armand Dezső. "Kolozsvári Czementes János könyve" [The book of János Cementes of Kolozsvár]. *Magyar Könyvszemle* 4 (1896): 276–301, 351–73.

Higley, Sarah L. *Hildegard of Bingen's Unknown Language: An Edition, Translation, and Discussion.* New York: Palgrave Macmillan, 2007.

István, Vadai. "Két XVII. századi titkosírás megfejtése" [Solution to two seventeenth-century ciphers]. In *Pálffy Kata leveleskönyve: Iratok Illésházy István bujdosásának történetéhez (1602–1606)* [Letter-book of Kata Pálfyy: Texts relevant for István Illésházy's exile], edited by Eötvös Péter, 183–89. Szeged: Scriptum, 1991.

Ivanov, Jordan. *Bogomilski knigi i legend.* Sofia: Pridvorna Pechatnica, 1925.

———. *Le livres et legendes bogomiles.* Paris: Maisonneuve-Larose, 1976.

Jenő, Pintér. *Magyar irodalomtörténete* [The history of literature in Hungary]. Budapest: Franklin, 1930.

Jerney, János. "Némi világosítások az ismeretlen jellemű rohonczi írott könyvre" [Some information concerning the unknown book of Rohoncz]. *Tudománytár* 8 (1844): 25–36.

Johannes de Caulibus. *Meditaciones vite Christi olim S. Bonaventurae attributae.* Edited by M. Stallings-Taney. Corpus Christianorum Continuatio Mediaevalis 153. Turnhout: Brepols, 1997.

Kahn, David. *The Codebreakers: The Comprehensive History of Secret Communication from Ancient Times to the Internet.* New York: Scribner, 1996.

———, ed. "Enemy Codes and Their Solution." *Cryptologia* 19 (1995): 166–97.

———. "The Man in the Iron Mask: Encore et Enfin, Cryptologically." *Cryptologia* 29 (2005): 43–49.

Kalmár, György. *Grammaticalische Regeln zur philosophischen oder allgemeinen Sprache, das ist, der Sprache aller Voelker Zeiten und Lebensarten.* Vienna: Kurzböck, 1774.

———. *Praecepta grammatica atque specimina linguae philosophicae, sive universalis.* Berlin: D. Iacobaeer, 1772.

———. *Precetti di grammatica per la lingua filosofica o sia universale propria per ogni genre di vita.* Rome: P. Giunchi, 1773.

Karttunen, Liisi. *Chiffres diplomatiques des nonces de Pologne vers la fin du XVIe siècle: Extraits des archives des princes Chigi à Rome.* Helsinki: Annales Academiae Scientiarum Fennice, 1911.

Keevak, Michael. *The Pretended Asian: George Psalmanazar's Eighteenth-Century Formosan Hoax.* Detroit: Wayne State University Press, 2004.

Kelecsényi, Ákos. "Egy magyar régiségkereskedő a 19. században: Literáti Nemes Sámuel (1794–1842)" [Sámuel Literáti Nemes: A Hungarian antique dealer in the nineteenth century]. In *Az Országos Széchényi Könyvtár Évkönyve, 1972* [Annual of the National Széchényi Library, 1972], 307–27. Budapest: OSZK, 1975.

———. "Literáti Nemes Sámuel útinaplója" [The travelogue of Sámuel Literáti Nemes]. In *Az Országos Széchényi Könyvtár Évkönyve, 1968–69* [Annual of the National Széchényi Library, 1968–69], 317–30. Budapest: OSZK, 1971.

Kelecsényi, Gábor. *Múltunk neves könyvgyű-jtői* [Famous book collectors of our past]. Budapest: Gondolat, 1988.

Kennedy, Gerry, and Rob Churchill. *The Voynich Manuscript*. London: Orion, 2004.

Kerekes, Dóra. *Diplomaták és kémek Konstantinápolyban* [Diplomats and spies in Constantinople]. Budapest: L'Harmattan, 2010.

———. "Hírszerzés a XVI–XVII században" [Intelligence in the sixteenth–seventeenth centuries]. *Irodalomismeret* 13 (2003): 63–70.

———. "Kémek Konstantinápolyban: A Habsburg információszerzés szervezete és működése (1683–1699)" [Spies in Constantinople, the system and functioning of Habsburg intelligence]. *Századok* 141 (2007): 1217–58.

King, John C. "A Reconstruction of the Key to Beale Cipher Number Two." *Cryptologia* 17 (1993): 305–17.

Király, Levente Zoltán. "Struktúrák a Rohonci-kódex szövegében: Helyzetjelentés egy amatőr kutatásról" [Structures in the text of the Rohonc Codex: A report on the findings of an amateur researcher]. *Theologiai Szemle* 54 (2011–12): 82–93.

Király, Levente Zoltán, and Gábor Tokai. "Cracking the Code of the Rohonc Codex." *Cryptologia* 42 (2018): 285–315.

Kircher, Athanasius. *Polygraphia nova et universalis ex combinatoria arte detecta*. Rome: Varesius, 1663.

Knowlson, James. "The Ideal Languages of Veiras, Foigny, and Tyssot de Patot." *Journal of the History of Ideas* 24 (1963): 269–78.

———. *Universal Language Schemes in England and France, 1600–1800*. Toronto: University of Toronto Press, 1975.

Koltai, András. *Batthyány Ádám és könyvtára* [Ádám Batthyányi and his library]. Budapest: OSZK; Szeged: Scriptum, 2002.

Koltai, Kornélia, ed. *Az Atyák mondásai: Péchi Simon kiadatlan rabbinikus írásai* [Sayings of the Fathers: Unpublished rabbinic writings of Simon Péchi]. Budapest: MTA, 1999.

Körmöczi, László. *A világnyelv kérdése és Kalmár György* [The problem of a world language and György Kalmár]. Nagykőrös: n.p., 1933.

Kovácsházi, István. *Pantographia (egyetemes írás)* [Pantography (universal writing)]. Budapest: Franklin Társulat, 1877.

Kruh, Louis. "A Basic Probe of the Beale Cipher as a Bamboozlement." *Cryptologia* 6 (1982): 378–82.

———. "The Beale Cipher as a Bamboozlement Part II." *Cryptologia* 12 (1988): 241–46.

Landini, Gabriel. "Evidence of Linguistic Structure in the Voynich Manuscript Using Spectral Analysis." *Cryptologia* 25 (2001): 275–95.

Láng, Benedek. "Invented Middle Ages in Nineteenth-Century Hungary: The Forgeries of Sámuel Literáti Nemes." In *Manufacturing a Past for the Present: Forgery and Authenticity in Medievalist Texts and Objects in Nineteenth-Century Europe*, edited by Patrick Geary and Klaniczay Gábor, 129–43. Leiden: Brill, 2015.

———. "People's Secrets: Towards a Social History of Early Modern Cryptography." *Sixteenth Century Journal* 45, no. 2 (2014): 291–308.

———. *Real Life Cryptology: Ciphers and Secrets in Early Modern Hungary*. Amsterdam: Amsterdam University Press, 2018.

———. *Unlocked Books: Manuscripts of Learned Magic in the Medieval Libraries of Central Europe*. University Park: Penn State University Press, 2008.

———. "Why Don't We Decipher an Outdated Cipher System? The Codex of Rohonc." *Cryptologia* 34 (2010): 115–44.

Leeuw, Karl de, and Hans van der Meer. "A Turning Grille from the Ancestral Castle of the Dutch Stadtholders." *Cryptologia* 19 (1995): 153–65.

Le Pape, Gilles. *Les écritures magiques: Aux sources du "Registre des 2400 noms" d'anges et d'archanges de Martines de Pasqually.* Milano: Arché, 2006.

Lewis, Rhodri. *Language, Mind and Nature: Artificial Languages in England from Bacon to Locke.* Cambridge: Cambridge University Press, 2007.

Locsmándi, Miklós. "A Rohonci Kódex: Egy rejtélyes középkori írás megfejtési kísérlete" [The Rohonc Codex: An attempt to solve a mysterious medieval writing]. *Turán* 7–8 (2004–5): 41–58.

Ludolf von Sachsen. *Vita Christi: Introductory Volume.* Analecta Cartusiana 24. Salzburg: Institut für Anglistik und Amerikanistik, Universität Salzburg, 2007.

Marton, Veronika. *I. András király korabeli imák: Az első magyar nyelvemlék* [Prayers from the time of King Andrew I: The first Hungarian piece of writing recovered]. Győr: Matrona, 2006.

Mavor, William Fordyce. *Universal Stenography; or, A New Complete System of Short Writing.* N.p.: Harrison, n.d.

Mead, G. R. S., ed. *Pistis Sophia: A Gnostic Gospel (with Extracts from the Books of the Saviour Appended).* London: The Theosophical Publishing Society, 1896.

Meister, Aloys. *Die Anfänge der modernen diplomatischen Geheimschrift.* Paderborn: Ferdinand Schöningh, 1902.

———. *Die Geheimschrift im dienste der päpstlichen Kurie von ihren Anfängen bis zum Ende des 16. Jahrhunderts.* Paderborn: Ferdinand Schöningh, 1906.

Molnár, Antal. *Katolikus missziók a hódolt Magyarországon, I. 1572–1647* [Catholic missions in Hungary during the oppression, part I, 1572–1647]. Budapest: Balassi, 2002.

———. *Lehetetlen küldetés? Jezsuiták Erdélyben és Felső-Magyarországon* [Mission impossible? Jesuits in Transylvania and Northern Hungary]. Budapest: L'Harmattan, 2009.

Mrayati, Mohamad, Yahya Meer Alam, and M. Hassan at-Tayyan, eds. *Al-Kindi's Treatise on Cryptanalysis.* Riyadh: King Faisal Center for Research and Islamic Studies, 2003.

———, eds. *Ibn ad-Durayhim's Treatise.* Riyadh: King Faisal Center for Research and Islamic Studies, 2004.

———, eds. *Ibn Adlan's Treatise.* Riyadh: King Faisal Center for Research and Islamic Studies, 2003.

———, eds. *Ibn Dunaynir's Book.* Riyadh: King Faisal Center for Research and Islamic Studies, 2005.

———, eds. *Three Treatises on Cryptanalysis of Poetry.* Riyadh: King Faisal Center for Research and Islamic Studies, 2006.

———, eds. *Two Treatises on Cryptanalysis.* Riyadh: King Faisal Center for Research and Islamic Studies, 2007.

Nagy, Ilona. *Apokrif evangéliumok, népkönyvek, folklór* [Apocryphal gospels, books of folk tradition, and folklife]. Budapest: L'Harmattan, 2001.

Némäti, Kálmán. *Rohonczi Codex Tantétel* [Statements about the Rohonc Codex]. Budapest: n.p., 1892.

The New Testament of Our Lord and Saviour Jesus Christ in Taylor's System of Short Hand as Improved by George Odell. London: G. Odell, 1843.

The New Testament of Our Lord and Saviour Jesus Christ, Printed in an Easy

Reporting Style of Phonography by Isaac Pitman. London: Frederick Pitman, 1886.

Nickell, Joe. *Detecting Forgery: Forensic Investigation of Documents*. Lexington: University Press of Kentucky, 1996.

———. *Pen, Ink, and Evidence: A Study of Writing and Writing Materials for the Penman, Collector, and Document Detective*. New Castle, DE: Oak Knoll Books, 1990.

———. *Real or Fake? Studies in Authentication*. Lexington: University Press of Kentucky, 2009.

Nyíri, Attila. "Megszólal 150 év után a Rohonci-kódex?" [Will the Rohonc Codex finally speak, after 150 years of silence?]. *Theologiai Szemle* 39 (1996): 91–98. Republished as "A Rohonci-kódexről" [About the Rohonc Codex]. *Turán* 7, no. 4 (2004): 85–92.

Pagels, Elaine. *The Gnostic Gospels*. New York: Vintage Books, 1979.

Pálffy, Géza. *The Kingdom of Hungary and the Habsburg Monarchy in the Sixteenth Century*. Boulder, CO: Center for Hungarian Studies and Publications, 2009.

Pasini, Luigi. "Delle scritture in cifra usate dalla Repubblica Veneta." In *Il Regio Archivio Generale di Venezia*, 291–328. Venezia: Pietro Naratovich, 1873.

Pellerey, Roberto. *Le lingue perfette nel secolo dell'utopia*. Rome: Laterza, 1992.

Pesic, Peter. "François Viète, Father of Modern Cryptanalysis—Two New Manuscripts." *Cryptologia* 21 (1997): 1–29.

Petercsák, Tivadar, and Mátyás Berecz, eds. *Információáramlás a magyar és török végvári rendszerben* [Information flow in the Hungarian and Turkish Military Zones]. Eger: Dobó István Vármúzeum, 1999.

Platania, Gaetano. "La Polonia nelle carte del cardinale Carlo Barberieni Protettore del regno." *Accademie e Biblioteche d'Italia* 56 (1988): 38–60.

Poe, Edgar Allan. "A Few Words on Secret Writing." *Graham's Magazine* 19 (1841): 33–38.

Principe, L. M. "Robert Boyle's Alchemical Secrecy: Codes, Ciphers and Concealments." *Ambix* 39, no. 2 (1992): 63–74.

Reeds, Jim. "William F. Friedman's Transcription of the Voynich Manuscript." *Cryptologia* 19 (1995): 1–22.

Rockinger, Ludwig von. "Über eine bayerische Sammlung von Schlüsseln zu Geheimschriften des sechzehnten Jahrhunderts." *Archivalische Zeitschrift* 3 (1892): 21–96.

Rossi, Paolo. *Clavis universalis: Arti della memoria e logica combinatoria da Lullo a Leibniz*. Bologna: il Mulino, 1983.

Rugg, Gordon. "An Elegant Hoax? A Possible Solution to the Voynich Manuscript." *Cryptologia* 28 (2004): 31–46.

Safford, L. F. "The Functions and Duties of the Cryptography Section, Naval Communications." *Cryptologia* 16 (1992): 265–81.

Sams, Eric. "Cryptanalysis and Historical Research." *Archivaria* 21 (1985–86): 87–97.

Sándor, Eckhardt. *Balassi Bálint összes művei* [The collected works of Bálint Balassi]. Budapest: Akadémiai Kiadó, 1951.

Sassoon, George T. "The Application of Sukhotin's Algorithm to Certain Non-English Languages." *Cryptologia* 16 (1992): 165–73.

Schinner, Andreas. "The Voynich Manuscript: Evidence of the Hoax Hypothesis." *Cryptologia* 31 (2007): 95–107.

Selenus, Gustavus. *Cryptomenytices et cryptographiae libri IX*. Luneburg: Sternen, 1624.

Seligman, H. "Un traité de déchiffrement du XVIIe siècle." *Revue des Bibliothèques et Archives de Belgique* 6 (1908): 1–19.

Singh, Mahesh Kumar. "Rohonci Kódex" [The Rohonc Codex]. With Hungarian translation by László Bárdi. *Turán* 7–8 (2004–5): 9–40.

———. "Rövid ismertető a brahmi ABC-ről" [A short introduction to the Brahmic Alphabet]. *Turán* 8, nos. 2–3 (2005): 133–38.

Singh, Simon. *The Code Book*. New York: Anchor, 2000.

Slaughter, Mary M. *Universal Languages and Scientific Taxonomy in the Seventeenth Century*. Cambridge: Cambridge University Press, 1982.

Speziali, Pierre. "Aspects de la cryptographie au XVI siècle." *Bibliothèque d'humanisme et Renaissance* 17 (1955): 188–206.

Stanley, Jeffrey Christopher. "To Read Images Not Words: Computer-Aided Analysis of the Handwriting in the Codex Seraphinianus." Master's thesis, North Carolina State University, 2010.

Stix, Franz. "Die Geheimschriftenschlüssel der Kabinetskanzlei des Kaiser." *Nachrichten von der Gesellschaft der Wissenschaften zu Göttingen, Philologisch-Historische Klasse* 2 (1936): 207–26 and 3 (1937): 61–70.

Strasser, Gerhard. "Die kryptographische Sammlung Herzog Augusts: Vom Quellenmaterial für seine Cryptomenytices zu einem Schwerpunkt in seiner Bibliothek." *Wolfenbütteler Beiträge* 5 (1982): 83–121.

———. *Lingua Universalis: Kryptologie und Theorie der Universalsprachen im 16. und 17. Jahrhundert*. Wiesbaden: Harrassowitz, 1988.

———. "The Noblest Cryptologist: Duke August the Younger of Brunswick-Luneburg (Gustavus Selenus) and His Cryptological Activities." *Cryptologia* 7 (1983): 193–217.

Szabó, Károly. "A régi hún-székely írásról" [About the Hunnic-Sekler writing]. *Budapesti Szemle* 2 (1866): 106–30.

Szaniszló, Zsigmond. *Naplói (1682–1711)* [Diaries, 1682–1711]. Edited by Torma Károly. *Történelmi Tár* 12 (1889): 230–69, 503–22, 708–27; 13 (1890): 77–101, 307–27, 493–510, 757–70; 14 (1891): 267–95.

Szerdahelyi, István. *Bábeltől a világnyelvig* [From Babel to a universal language]. Budapest: Gondolat, 1977.

Szilády, Áron, ed. *Sermones Dominicales: Két XV. századból származó magyar glosszás latin codex* [Sermones Dominicales: Two Latin codices with Hungarian glossaries from the fifteenth century]. 2 vols. Budapest: MTA, 1910.

Szinnyei, József. *Magyar irók élete és munkái* [The lives and works of Hungarian writers]. Vol. 9. Budapest: Hornyánszky Viktor, 1903.

Tenschert, Heribert, and Ina Nettekoven. *Horae B.M.V. 158 Stundenbuchdrucke der Sammung Bibermühle, 1490–1550*. 3 vols. Wiesbaden: Harrassowitz, 2003.

Thorndike, Lynn. *History of Magic and Experimental Science*. New York: Columbia University Press, 1923–58.

Tokai, Gábor. "Az első lépések a Rohonci-kódex megfejtéséhez" [The first steps toward an undeciphering of the Rohonc Codex]. *Élet és Tudomány* 65 (2010): 1675–78; 66 (2011): 50–53.

———. "A Rohonci-kódex művészettörténész szemmel" [The Rohonc Codex through the eyes of an art historian]. *Élet és Tudomány* 65 (2010): 938–40, 1004–6, 1104–6.

Tóth, Béla. *Magyar ritkaságok* [Hungarian rarities]. Budapest: Athenaeum, 1899.

Trithemius, Johannes. *Polygraphiae libri sex.* Oppenheim: Haselberg de Aia, 1518.

———. *Steganographia: Ars per occultam scripturam.* Frankfurt: Becker, 1606.

Turdeanu, Emil. "Apocryphes bogomiles et apocryphes pseudo-bogomiles." *Revue de l'Histoire des Religions* 138 (1950): 22–52, 176–218.

———. *Apocryphes slaves et roumains de l'Ancien Testament.* Leiden: Brill, 1981.

Tusor, Péter. "Pázmány bíboros olasz rejtjelkulcsa" [The Italian cipher key of Cardinal Pázmány]. *Hadtörténelmi közlemények* 116 (2003): 535–81.

Ungureanu. Dan. "Nu trageti in ambulanta." *Observator Cultural* 167 (May 6–12, 2003). https://www.observator cultural.rov/articol/nu-trageti-in -ambulanta.

United States Department of the Army. *Basic Cryptanalysis.* Field Manual 34-40-2. Washington, DC, 1990. https://fas.org/irp/doddir/army /fm34-40-2.

Varga, Csaba. "A Rohonczi Kódex M K Singh-féle olvasatának ellenőrzése" [Checking M K Singh's transcript of the Rohonc Codex]. *Turán* 9, nos. 2–3 (2005): 198–202.

Várkonyi, Ágnes R. "Az elveszett idő: Zrínyi Miklós nádori emlékirata?" [Time lost: The memoirs of palatine Miklós Zrínyi?]. *Hadtörténeti Közlemények* 113 (2000): 269–328.

———. *Thaly Kálmán és történetírása* [Kálmán Thaly and his history writing]. Budapest: Akadémiai Kiadó, 1961.

———. "Történészvita Zrínyiről 1868-ban" [Historians debating Zrínyi in 1868]. In *Tanulmányok Pölöskei Ferenc köszöntésére* [Studies celebrating Ferenc Pölöskei], edited by Jenő Gergely, 627–40. Budapest: ELTE BTK, 2000.

Vigenère, Blaise de. *Traicte des Chiffres.* Paris: Abel l'Angelier, 1586.

Viterbo, Emanuele. "The Ciphered Autobiography of a 19th Century Egyptologist." *Cryptologia* 22 (1998): 231–43.

Wehli, Tünde, and Loránd Zentai, eds. *Biblia pauperum és előtte a Vita et passio Christi képei a Szépművészeti Muzeum kódexében* [The illustrations of the Biblia pauperum and the Vita et passio preceding it in the codex of the Museum of Fine Arts of Budapest]. Budapest: Európa, 1988.

Westfall, Richard S. "Short-Writing and the State of Newton's Conscience, 1662." *Notes and Records of the Royal Society of London* 18 (1963): 10–16.

Wilkins, John. *Mercury, or The Secret and Swift Messenger.* London: Baldwin, 1694.

Willis, John. *The Art of Stenographie, teaching by plaine and certaine rules.* London: Cuthbert Burbie, 1602.

Winkel, Brian. "Poe Challenge Cipher Finally Broken." *Cryptologia* 1 (1977): 93–96.

———. "Poe Challenge Cipher Solutions." *Cryptologia* 1 (1977): 318–25.

Wrixon, Fred B. *Codes, Ciphers, and Other Cryptic and Clandestine Communication.* New York: Black Dog, 1998.

Agrippa, Cornelius, 115, 119
Alberti, Leon Battista, 67
Al-Kindi, 65, 87
Al-Qalqashandi, 65–66
American Declaration of Independence, 17
Andrew I, King, 12, 15
Arabic language, 55, 59, 66, 100
Argenti, Giovanni Battista, 66
Argenti, Matteo, 66
Armenian language, 55, 60, 66

Babbage, Charles, 68
Bacon, Francis, 108, 120
Bacon, Roger, 17
Baghdad, 65
Balassi, Bálint, 77
Banská Štiavnica codex,10, 13
Barabbas, 89–90, 94
Barmby, Henry, 100
Batthyány family 4, 6, 8, 14, 37–38, 108
Batthyány, Gusztáv, 4
Bazeries, Étienne, 40
Beale ciphers, 3, 16–17, 134
Beck, Cave, 103, 106
Beinecke Library at Yale University, 18
Bernonville, Pierre de, 106
Bible, 54–58, 60, 94, 100, 102, 106
    Abraham preparing to sacrifice Isaac, 137
    Annunciation, 42, 45, 53, 138
    Ascension, 42, 140
    Christ arrested in the Garden of
        Gethsemane, 30, 138
    Christ bearing the cross, 42, 139
    Christ before Pilate, 30, 42, 47, 52, 55,
        123, 139
    Christ entering Jerusalem, 30, 42, 44,
        54, 138
    Cleansing of the Temple, 44, 54
    Crucifixion, 30, 42, 46, 138, 139
    Doubting Thomas, 140
    Epiphany, the Adoration of the Magi 30,
        42, 45, 138

    Flagellation of Christ, 139
    INRI, 52, 94, 96, 138, 139
    Moses with the Tablets of Law on Mount
        Sinai, 44, 137
    Noli me tangere, 30, 48, 140
    Resurrection, 42, 49, 52, 55, 140
    Sacrifice of a pigeon in the Temple, 45,
        54, 138
    Transfiguration, 46, 138
    Wedding at Cana, 143
Biblia pauperum, 57
Bibliothèque nationale, 6
Blaki state, 27–28
Bogomil sect, 56, 60
Bolyai, János, 103, 109, 112
Bonaventure, Saint, 57
book cipher, 16
Book of Hours, 57, 60
Borges, Jorge Luis, 105, 113
Boyle, Robert, 77
Bright, Timothy, 98, 101
Brosses, Charles de, 106
Bruno, Giordano, 104
Briquet, Charles-Moïse, 35
Buda, 10, 14, 24, 72–73
Byzantium, 29

Campanella, Tommaso, 104
Caesar, Julius, 64
Cementes, Johannes, 77
Chadwick, John, 84
Champagnolles, D. P., 120
Charlemagne, 64
Charles I, King, 120
Chatar sect, 56, 60
Cherokee, 108
Chicago, 7
codebreaking. See cryptanalysis
Codex Seraphinianus, 116–17
Comenius, John Amos, 104
Constantinople, 73, 109
Coptic language, 55

Cospi, Antonio Maria, 81
Cree, 108
Creed, 19, 112
cryptanalysis
    frequency analysis, 34, 63–67, 69–70,
        82–83, 87–96, 120–21
    identification of isologues, 90–91, 94,
        122, 124
    probable-word analysis, 70, 90, 93–94,
    Sukhotin's algorithm, 89
    word-pattern analysis, 89, 95, 121
    vowel analysis, 6, 67, 82–83, 89, 92–93,
        95, 121
cryptophasia, 39
Czech language, 55

Dalgarno, George, 103, 106
Dead Sea Scrolls, 55
Dee, John, 17, 115, 119–20
Della Porta, Giambattista, 67
des Vallées, 120
Descartes, René, 103
Dumas, Alexandre, 72

Eastern Orthodoxy, 72
Eco, Umberto, 103
Egypt, 55,
Egyptian hieroglyphs, 2, 51, 127
Enăchiuc, Viorica, 26–32, 34, 39
Enemy Codes and Their Solution, 86
English Civil War, 99
Enigma code, 80
entropy, 67–69, 87, 92, 133
Esperanto language, 106, 112
espionage, 76
European Skeptics Congress, 36

Falconer, John, 67
Foigny, Gabriel de, 115
Fontana, Giovanni, 77
forgery, 5, 8, 12, 14, 22, 24–25
Formosa (Taiwan), 115–16
founding charter of the Tihany Abbey, 12
frequency analysis. See cryptanalysis
Friedman, William, 4, 84–85, 87, 92, 134
    Military Cryptanalysis, 84

Galilei, Galileo, 77, 111

Gárdonyi, Géza, 33, 98
Gáti, István, 109, 111–12
Gelle prayer book, 19–21
German language, 1, 38, 51, 61, 64, 73, 86–87,
    106, 109, 116–17
Gesta Hungarorum, 28
Gilicze, Gábor, 33
Gospel
    of John, 54
    of Matthew, 54
    of Pilate, 55
    of Thomas, 55
Gospels
    apocryphal, 54–61
    canonized, 42–43, 52, 54, 94, 100, 128–29,
        138
    Gnostic, 54–55, 60
grille, 87
Guénin, Louis-Prosper, 99
Guldin, Pierre, 105
Gyürk, Ottó, 33

Habsburg rulers, 9, 73–76, 78, 82
Hail Mary, 19, 53
Haller, Gábor, 78
Hawkes, James, 100
Henry IV, French king, 82
Herod, 30
Hildegard of Bingen, Saint, 115, 119
Hitt, Parker, 84
hoax theory, 2–3, 6–7, 15–22, 36, 39, 58, 96,
    103, 114–18, 132–34
Homer, 120
homophonic cipher, 69–75, 77, 79–83, 87,
    89–96, 99, 124, 133
Humphreys, James, 100
Hunfalvy, Pál, 24
Hungarian Academy of Sciences, 4–5, 9,
    19–21, 24–25
Hungarian language, 5, 8, 10–13, 15, 19,
    23–26, 32, 34, 38–39, 51, 61, 64, 75–77,
    88–89, 104, 109, 112, 117, 130, 132
Hunor, 28
Hügel, Delia, 131

Ibn ad-Durayhim, 65
Ibn Adlan, 65
Illésházy, István, 78

Indus Valley script, 118
International Congress on Medieval Studies
    in Kalamazoo, 5, 10
iron gall ink, 37

Janus Pannonius, 8
Jerney, János, 10, 23, 37, 134
Jews, 17, 94
Jireček, Josef, 24
Johnson, Reverend, 120–21
Joseph, 30, 45, 138
Judaizing sects, 59
Jülg, Bernath, 24

Kabbalah, 103, 104
Kájoni, János, 19
Kalmár, György, 103, 106–11, 121
Kelley, Edward, 17, 115, 120
Khazars, 17, 24
Kinder, Philip, 120
Király, Levente Zoltán, 34, 53–54, 58, 125–31,
    134, 137
Kircher, Athanasius, 103, 104, 106, 127–28
Kovácsházi, István, 109, 112–13
Krakow, 75

L'art de deschiffrer, 83
Latin language, 10–12, 18–19, 26–30, 38, 51,
    62, 65, 68, 73, 76, 82, 92–93, 98–102,
    105–6, 109, 111, 115, 117
Le Clercq, Chrétien, 108
le Maire, Jean, 120
Leibniz, Gottfried Wilhelm, 103–5, 109, 111
Levi, Simeone, 41, 90, 98
Liber runarum, 76
Linear A writing, 2–3
Linear B writing, 2–3, 84, 118, 129
Literáti Nemes, Sámuel, 8–15, 22, 37
Locsmándi, Miklós, 33–34
Lodwick, Francis, 103, 106–7, 120
Lord's Prayer, 19, 100, 106–7, 112–13, 116, 129
Louis XIV, 40, 70, 72, 75
Ludolf the Carthusian, 57
Ludovicus de Berlaymont, 63
Lullus, Raimundus, 103–4

Macpherson, James, 9
Maimieux, Joseph de, 103, 106

Martian language, 116–19
Marton, Veronika, 12
Mary, mother of Christ, 30, 45, 138, 143
Mary Magdalene, 48, 140
Máté, Imre, 74
Mátray, Gábor, 10
Matthias Corvinus, King, 9
Mersenne, Marin, 103, 105
Milan, 71
Monoalphabetic cipher, 21, 62–69, 76–79,
    82–83, 87, 90, 92–93, 115, 120
More, Thomas, 115
Moses, 44, 131, 137
Motmann, Cornelius Heinrich, 73
Munkácsy, Mihály, 24
Müller, Alois, 24

Nag Hammadi scrolls, 54–55
Navajo language, 86
Némäti, Kálmán, 24–25, 34, 38
Neurath, Otto, 112
Newton, Isaac, 99, 103, 106, 109, 111
Nickell, Joe, 36–37
nomenclators, 70–71, 74–75, 80, 82, 87, 92,
    96, 121
nullities, 70–71 79, 80, 83, 93, 96, 99, 101
Nyíri, Attila, 25–26, 32, 34, 39

Oancea, Marius-Adrian, 130–31
Oghuz people, 28, 30
Old Church Slavonic language, 60
old-Bulgarian language, 55
Ossian poems, 9
Ottoman Empire, 51, 59–60, 72–74, 76

Paris, 6–7, 24, 99
Patot, Simon Tyssot de, 115
Pázmány, Péter, 73–74
Pechenegs, 28, 30
Pelling, Nick, 130
Pepys, Samuel, 99, 101
Phaistos Disk, 2–3, 14
pigpen, 78
Pilate, 30, 42, 47, 52, 54–57, 93–94, 123, 131,
    139
Pistis Sophia, 55
Pitman, Isaac, 100
Poe, Edgar Allan, 85

polyalphabetic cipher, 67–69, 87–88, 92, 133
Postel, Guillaume, 104
Potter, Beatrix, 81
probable word method. *See* cryptanalysis
Psalmanazar, George, 106, 115–16, 120

Ragusa, 73
Rákóczi, Ferenc II, 73, 74–75
Rákóczi's War of Independence, 9, 74
Rechnitz, 4, 130
Reformation, 56, 72–73
Romanian language, 26–29, 51, 61
Rome, 4, 73
Rongorongo, 2, 118
Rosetta Stone, 51
Rosicrucian Brotherhood, 1
Royal Society, 116
Rudolf II, Emperor, 17

Sainte-Geneviève Library, Paris, 6, 99–100
Sambucus, Johannes, 9
Sanskrit language, 31–33
Seidel, Martin, 59
Selenus, Gustavus – August of
        Braunschweig, 67, 69, 92
Serafini, Luigi, 116–18
Shelton, Thomas, 98–99
shorthand. *See* stenography
Sieniawska, Elżbieta Helena, 75
Singh, Mahesh Kumar, 31–33, 39
Singh, Simon, 63
Skytte, Benedict, 120
Smith, Hélène, 116–20
Smith, John, 99
Songs of Christ, 57
Southern Slavic language, 51, 76
steganography, 69
stenography, 39–40, 94, 96–101, 133
Sukhotin's algorithm. *See* cryptanalysis
Szabó, Károly, 10, 14, 19
Szaniszló, Zsigmond, 78
Széchényi, Ferenc, 9–11

Tartars, 23
Ten Commandments, 112, 116, 19

Thaly, Kálmán, 9
Thököly, Imre, 74
Tiro, Marcus Tullius, 98–99
Tironian notes, 98–99
Tisza, river, 27–29
Tokai, Gábor, 34, 53–54, 58, 124–31, 134, 137
Toldoth Jeschu, 59
Toldy, Ferenc, 24
Tolkien, J. R. R., 115
Transylvania, 9, 13, 28, 51, 59, 73, 75–76,
        78–79
Trithemius, Johannes, 67, 69, 103
Turán, journal, 31–33
Turkish language, 51, 59–61, 73, 94
Turks, 24, 59, 72, 75, 79
Tusor, Péter, 74

Vatican, 7
Veiras, Denis, 115
Venice, 9, 66, 73
Vienna, 73, 112
Vienna Circle, 112, 114
Viète, François, 82–83
Vigenère, Blaise, 67, 69
Villa Mondragone, 4
Vita Christi, 56
Vitéz, Johannes, 8
Vlad, emperor, 27, 29–30
Volapük language, 106, 112
Voltaire, 72
Voynich manuscript, 2–6, 17–18, 117–18, 132
Voynich, Wilfrid, 3, 17
Vulgar Latin language, 26–30

Wallis, John, 97
Warsaw, 75
Webster, John, 108
Wesselényi, Ferenc, Palatine, 74
Weston, Janes, 100
Wilkins, John, 97, 103, 105–9, 111, 119–21
Willis, John, 97–98, 101
Wit-spell, 121